SPACE CHEMISTRY

*Ann Arbor
Science
Library*

SPACE CHEMISTRY

by Paul W. Merrill

ANN ARBOR

THE UNIVERSITY OF MICHIGAN PRESS

Second printing 1966

Copyright © by The University of Michigan 1963
All rights reserved
Library of Congress Catalog Card No. 60-15776
Published in the United States of America by
The University of Michigan Press and simultaneously
in Rexdale, Canada, by Ambassador Books Limited
Manufactured in the United States of America

PREFACE

A century ago laboratory chemistry was already well developed but nothing at all was known of the matter composing the heavenly bodies. Indeed, Auguste Comte, a famous and able philosopher, had stated categorically that the chemical composition of the stars must remain forever outside the domain of human knowledge. Comte's mind was encyclopedic but he did not foresee the possibilities of the spectroscope as a tool of analytical chemistry.

A great new era in astronomy began in 1859 when Kirchhoff and Bunsen found that various gases can be easily and positively identified by a detailed study of the light they emit or absorb. This optical method was quickly applied with great success first to the sun and then to the distant fixed stars. The new chemical and physical studies of the heavenly bodies were soon receiving as much attention as the older form of astronomy which dealt with their brightness, direction, and motion.

A comprehensive chart of the universe is hard to achieve. This is because we have no outside vantage point from which various parts appear in their true proportions. Instead, we are on the inside looking out into a glorious but bewildering array of planets, stars, nebulae, and galaxies. Except for objects in our relatively tiny solar system and for a few of the nearest stars, astronomical distances are so great that we cannot measure them directly by any means now at our command. Our knowledge of distances and of the general plan of the universe rests on complex statistics gathered with great labor and not easily interpreted. Our own Milky Way system is but one of billions of galaxies of stars scattered throughout observable space —and even this major fact was established less than fifty years ago. Now the general map of the observable universe is quite plainly drawn, but the scale is still subject to revision.

In studying the chemistry of the stars, on the other hand, the situation is much more favorable. Here the spectroscope, the instrument used for chemical analysis, is so powerful that distance causes little difficulty. Its power is illustrated by the fact that in the laboratory, where a chemist can handle material and put it in a test tube, he often prefers to use the spectroscope. Instead of subjecting the sample to various reagents he may put some of it into an electric arc and gather information in exactly the same way as an astronomer. Both must collect some light—one from a laboratory source, the other from a star—and send it into a spectroscope. Light waves can reveal an immense amount of chemical and physical data. Although the reports become fainter as the light spreads out from the source, they remain distinct and accurate, and even from vast distances, when recorded on a photographic plate, are still legible. The spectroscope is used to separate the colors in a light beam gathered, in the case of stars, by a telescope, and to lay them out in order in a spectrum. After the spectrum of a star is photographed, a glance at the developed plate often brings, to the trained eye, evidence of half a dozen chemical elements. Measurement, or comparison with spectra previously analyzed, may reveal many more. Hundreds of thousands of stars have undergone partial analysis. A large part of modern astronomy might appropriately be called astrochemistry.

I hope, however, that the assertion by Miss Agnes Clerke, brilliant English writer on astronomy, will always remain true. She wrote: "Custom can never blunt the wonder with which we must regard the achievement of compelling rays emanating from a source devoid of sensible magnitude through immeasurable distance [she was referring to the pointlike image of a star] to reveal, by their distinctive qualities, the composition of that source."

Modern astronomy is thus closely related to laboratory research in chemistry and physics. I hope the underlying unity of science will be illustrated and emphasized in the following pages in which I have tried to outline as simply as possible our present extensive knowledge of the chemical composition of our complex universe.

Excellent nontechnical descriptions of the spectroscope and its work will be found in two recent books in this series: (1) *Light: Visible and Invisible* by Eduard Ruechhardt, Chapter IX, "Light Dispersion," and Chapter XII, "Spectral Lines, and What They Tell"; (2) *The Stars* by W. Kruse and W. Dieckvoss, Part One, III, "Color (The Spectrum)"; both volumes by The University of Michigan Press. A reader who would like further knowledge of the chemical composition of stars might consult the introduction and the sections on the important astronomical elements hydrogen, helium, carbon, oxygen, calcium, and iron in my monograph *Lines of the Chemical Elements in Astronomical Spectra* (Carnegie Institution of Washington, 1956).

I am indebted to Dr. Alfred H. Joy, editor of the *Leaflets of the Astronomical Society of the Pacific*, for permission to use some material from the *Leaflets*.

P. W. M.

CONTENTS

I.   Uniformity and Diversity of Nature   13
II. Historical   18
III. Individual Chemical Elements   34
IV. Molecules   62
V.   Chemical Survey of the Universe:
     Solar System   69
VI. Chemical Survey of the Universe:
     Stars and Nebulae   107
VII. Atoms, Energy, and Evolution   138
VIII. Rockets, Satellites, and
     Space Chemistry   151
     Index   163

SPACE CHEMISTRY

The 200-inch Hale Telescope at Palomar Mountain, California.

I. UNIFORMITY AND DIVERSITY OF NATURE

Ancient scientists commonly believed that there were two sets of physical principles, one for the earth and another for the heavens. Within the past century, however, the principle of the uniformity of nature has been thoroughly established. This principle proclaims that the elementary physical particles that compose all atoms and molecules are always and everywhere the same, and that matter, electricity, gravity, and radiation all obey laws which are independent of time or position in space.

This principle of uniformity has been confirmed by many precise observations. For example, spectroscopic observations show that atoms of various elements such as hydrogen and calcium give out the same wavelengths of light—and therefore must have exactly the same internal structure—wherever they are found, whether it be in laboratories all over the earth, or throughout our vast Milky Way system, or in the most remote galaxies man has yet observed. Scientists everywhere have adopted this principle as a practical working hypothesis. More than two centuries ago Sir Isaac Newton, in stating his laws of motion and of gravitation, made no restrictions on time or place.

He made no distinction between heavenly bodies and those on earth. His first law of motion reads: "Every body continues in a state of rest or of uniform motion in a straight line unless it is compelled to change that state by a force impressed upon it." Similarly his law of gravitation was expressed in the sweeping statement: "Every particle of matter in the universe attracts every other particle with a force that varies inversely as the square of the distance between them and directly as the product of their masses."

Harlow Shapley and Edwin Hubble, the great modern explorers of the universe, based their conclusions concerning the size of our galaxy and the distances to other galaxies on a firm belief in the uniformity of nature as exemplified in the size, brightness, and behavior of certain kinds of stars which they used as indicators of distance, and Niels Bohr in setting up his marvelous quantum theory of the hydrogen atom has assumed confidently that this kind of atom wherever or whenever found has exactly the same internal structure.

Because of the uniformity of nature, knowledge gained in either the earth or the heavens is universal. Thus we use our knowledge of earth to investigate the stars, and we use astronomical observations to add to our understanding of the properties of atoms and the laws of nature.

One possible exception to the uniformity of atoms is "antimatter." This would consist of atoms with electric charges exactly opposite to those of our usual atoms. On earth the central nuclei of all ordinary atoms have positive electric charges which, in complete atoms, are neutralized by outer electrons with negative charges. In antimatter, the nuclei would be negative and the surrounding electrons positive. In 1931, electrons with positive charges, called "positrons," made by cosmic rays were observed by Carl Anderson in his laboratory in Pasadena, California. Then in 1955 a team of nuclear physicists headed by Emilio Segrè, working with the cyclotron in Berkeley, California, announced that they had observed negative protons, particles just like the nuclei of hydrogen atoms except that they had negative electric charges instead of the usual positive ones. These were ejected from a copper target bombarded by ordinary positive protons having the enormous energy of 6.2 billion electron volts. In our laboratories these particles with "reversed" electric charges last only a very small fraction of a second; they disappear with a flash of radiation when they meet their counterparts with regular charges.

The chemical properties of antimatter and the kinds of light it can emit or absorb when gaseous would be identical with those of ordinary matter. At a distance from the other, each would go its own stable course without trouble. But if matter and antimatter should meet, both would instantly disappear in a tremendous flash of radiation of very short wavelength. Thus the only way astronomers could detect antimatter would be through the energy released if it collided with normal matter. In a galaxy such as our own Milky Way system the materials are slowly but continually being mixed. Hence it is improbable that any considerable amount of antimatter exists in our galaxy, or that the two kinds of matter occur together in any one galaxy. But some distant isolated galaxy might, as far as we know, consist wholly of antimatter. This possibility is important but does not affect present-day studies of cosmic chemistry.

Uniformity of nature is God's greatest material gift to mankind. As Alexander Pope expressed it, "Order is Heaven's first law." Without order in nature, men might be automatons or clowns; past experience would supply no basis for future action, and men could not be rational beings capable of intelligent planning.

Nature is uniform but at the same time it is wonderfully diverse. The explanation of this seeming paradox is the task of science. Although the task is still incomplete, inspiring progress has been made. We now realize that all known substances, hundreds of thousands of them, are made of one or more of 103 chemical elements, and that these elements, in turn, are formed of only three elementary particles—protons (with positive electric charges), neutrons (no charges), and electrons (negative charges), each "uniform" in space and time.

The behavior of matter from subatomic particles, through atoms and molecules, all the way up to planets, stars, and galaxies is governed by various forces. These forces are constantly changing and their interplay is most complex. But only a few kinds of forces are involved. For small bodies, the most important are: (1) the powerful forces, not yet well understood, that hold together the nuclei of atoms; and (2) electric and magnetic forces whose effects can be accurately computed by the principles of physics. In the astronomical realm, a third force, gravitation, prevails, but it is now being recognized that electric and magnetic forces play a larger part than was formerly realized.

Thus the kinds of elementary particles and the kinds of forces are few, and the laws they obey are uniform. But the variety of properties and phenomena arising in the range of actual circumstances is astonishing. As a relatively simple example, consider the various circumstances under which oxygen atoms occur in nature. In the rarefied gaseous nebulae, oxygen occurs as atoms, each individual atom being alone. An atom may occasionally collide with other atoms, but it has no steady partners. In the earth's atmosphere, almost all the oxygen is in molecules, mostly in combinations of two atoms (O_2), a little in combinations of three atoms (O_3) called ozone. In water, an atom of oxygen is firmly bound to two of hydrogen (H_2O). In minerals in the crust of the earth, oxygen atoms are bound with atoms of metals to form oxides, and one or more oxygen atoms are present in many complicated substances, organic and inorganic, having several other kinds of atoms. These compounds differ enormously in their properties; nevertheless, the oxygen atoms are identical in them all. We must distinguish carefully between the fundamental, inherent properties of atoms and the circumstances in which atoms are found in various times and places.

The diversity of nature that we see about us is God's second material gift to mankind. The vast range of possibilities we are now just beginning to utilize was built into the foundations of the universe. Without diversity, men might become bored and as dull as oxen or snails; they could scarcely be intelligent human beings facing stimulating and ever-expanding possibilities of knowledge and of action. James Clerk Maxwell, the great nineteenth-century English physicist, once remarked: "It is a universal condition of the enjoyable that the mind must believe in the existence of law and yet have a mystery to move in."

II. HISTORICAL

Astronomy and chemistry, sciences whose roots are hidden in antiquity, developed in isolation from one another, in the darkness of ignorance, hearsay, and fiction. Early folk tales and mythologies developed into the pseudosciences of astrology and alchemy. Learning was for centuries a blend of observation, imagination, and superstition.

Although many of the ideas were worthless, a solid foundation of fact was nonetheless slowly being laid. Sundials and similar devices were used as early as 1500 B.C. to measure the direction to the sun. About 150 B.C. in Greece, Hipparchus began the systematic cataloguing of star positions, and 350 years later Ptolemy in Egypt prepared the *Almagest*, an important treatise on the motions of stars and planets. Until the sixteenth century astronomers believed his theory that the earth was the center about which all heavenly bodies revolved. But early in that century, Copernicus caused a major revolution in thought by teaching that the earth rotates on its axis and that the sun, not the earth, is the center of the system of planets. From then until now knowledge of the structure of the universe has steadily increased.

FIG. 1. Two of Galileo's telescopes and the lens (below) with which in 1610 he discovered the moons of Jupiter and observed the rings of Saturn and spots on the sun. Telescopic observations opened a new era in astronomy.

Because the universe as a whole is so large and seems to change so little, almost no progress in understanding its large-scale structure was made until early in this century when Harlow Shapley discovered a new way of estimating the distances of the globular star clusters. This discovery enabled him to outline roughly the domain of the Milky Way system. Shortly thereafter Edwin P. Hubble demonstrated by similar means that the mysterious spiral "nebulae" are actually independent galaxies far beyond the limits of our own local Milky Way system.

When Galileo directed his telescope toward the heavens in 1610 (Fig. 1) much of modern astronomy was begun.

For two and one-half centuries improved telescopes presented to the human eye ever brighter and larger images of the heavenly bodies. Then the photographic dry plate began to replace the human retina as the direct receiver of the images.

Astronomy and physics were brought together by Sir Isaac Newton, who applied the concepts of mass, inertia, and force, and his own inverse-square law of universal gravitation to astronomical motions. Since then, astronomy and physics have remained closely bound—to their mutual benefit.

The first connection between astronomy and chemistry was purely formal. In ancient days the Chaldeans identified the seven major bodies in the solar system with the seven metals then known. They bore the same names:

Sun	Gold
Moon	Silver
Mercury	Mercury
Venus	Copper
Mars	Iron
Jupiter	Tin
Saturn	Lead

The only such correspondence to survive in the English language is Mercury, but the idea of associating planets with chemical elements has persisted. In 1789 a newly discovered element was named uranium after the planet Uranus, and later, selenium and helium were named, respectively, after the Greek words for moon and sun. And only a few years ago the "artificial" elements just beyond uranium in the periodic table were named neptunium and plutonium after the outermost known planets.

In addition to his basic contributions to mechanics, Newton made a fundamental discovery of far-reaching consequence when he experimented with the effect of a glass prism on a beam of sunlight (Fig. 2). When the beam was deflected by the prism, he saw all the colors of the rain-

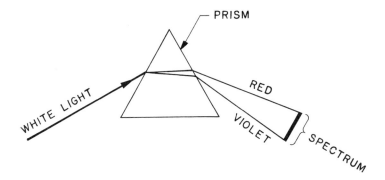

FIG. 2. Using a glass prism Sir Isaac Newton discovered that light of different colors is refracted by different amounts. White light or sunlight is made up of various colors which may be separated by a prism.

bow. Previous observers had been aware of the phenomenon but supposed that the colors were somehow made by the prism. Newton was the first to realize that they all were actually contained in the original beam of sunlight and that the prism did nothing more than separate the various colors by sending them in slightly different directions so that they could be seen side by side in a neat row which we now call a spectrum.

To Newton the band of colors in the spectrum seemed continuous, extending without any noticeable gaps or irregularities from violet at one end through blue, green, yellow, and orange to red at the other end. Detailed study of the spectra of light sources in the laboratory and in the sky has been made possible by continued improvements in spectroscopic equipment. The light to be examined is fed into a spectrograph which may be mounted on the solid ground or attached to the moving telescope (Fig. 3).

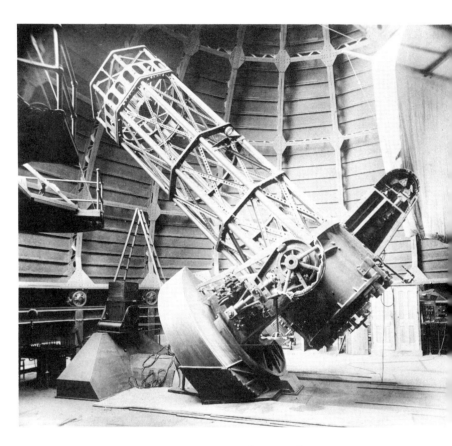

FIG. 3. For observing the spectra of stars with the 60-inch Mount Wilson reflector the box containing the spectrograph is attached just outside the cassegrain focus. In the box are the collimating lens which renders the light-rays parallel, the prism or prisms, the lens which forms the image of the slit in different colors, and the photographic plate. The telescope follows the star as it moves from east to west.

FIG. 4. The relative positions of the chief lines of the solar spectrum with their Fraunhofer designations and chemical identifications.

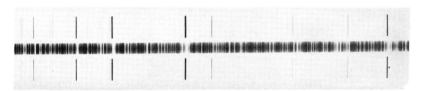

FIG. 5. Spectrum of the bright star Arcturus showing the dark absorption lines of many elements whose wavelengths are between 4330 A and 4425 A. Above and below the star spectrum are the emission lines of the iron arc which are useful for comparison. The same iron lines may be recognized in the star spectrum but in absorption. Most of the other star lines are due to titanium, vanadium, and cobalt. (Negative.)

The first observer to notice breaks in the spectrum of sunlight was W. H. Wollaston in 1802. A more detailed study of these narrow gaps—we now call them *dark lines*— was made about 1814 by Joseph von Fraunhofer who actually drew an excellent map of the lines and assigned letters to the strongest of them (Figs. 4 and 44). Although he found similar markings in the spectra of a few bright stars (Fig. 5), neither he nor anyone else grasped their significance. The correct explanation of the gaps was delayed until 1859 when the German scientists Gustav Robert Kirchhoff and Robert Wilhelm von Bunsen* together set

*This team of scientists deserves mention. Kirchhoff was a theoretician who made many important contributions to mathematical physics. Bunsen was an ingenious laboratory chemist. In 1855 he devised the famous "Bunsen burner." The chemist Wilhelm Ostwald wrote: "Every phenomenon embraced for him an endless diversity of factors, and in the yellow flame of an ordinary alcohol lamp whose wick was sprinkled with salt, he saw the possibilities of accomplishing the chemical analysis of the most distant stars."

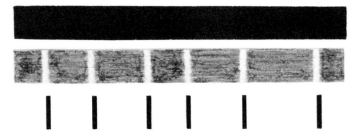

FIG. 6. A continuous spectrum is an uninterrupted series of images of the slit of all wavelengths. Such a spectrum comes from a hot solid body. Absorption lines have less intensity than the continuous spectrum on either side of them. Emission lines have greater intensity than the background. (Negative.)

forth clearly the chemical principles of spectroscopy. These men recognized, and for the first time stated clearly, this basic principle: If it is in the form of a gas, and is made to give out light, each chemical element will produce a characteristic set of bright spectrum lines quite different from that of any other element. If white light (which has a smooth continuous spectrum) is passed through the gas, the same characteristic lines will be absorbed and will become visible as dark lines (Fig. 6). This knowledge made chemical analysis of the heavenly bodies possible only a few years after Auguste Comte had declared it forever impossible.

From this first significant contact between astronomy and chemistry, man has developed a most wonderful teletype system, usually called spectroscopy. By means of it we receive directly from the stars, in a code not too difficult to interpret, reams of chemical data.

Because this code is based on spectrum lines we must understand what they really are. An ordinary beam of light, whether from the sun or from the filament of an electric lamp, contains many colors but they travel together in such a way that the eye sees them only as combined into "white" light. (White is not an individual color; it is always a combination of two or more colors.) An optical instrument called the spectroscope separates the various colors, laying them side by side in an orderly row—a spectrum—where each is free from the superposition of other colors.

Color is the sensation which the eye reports when it receives a beam of light. The objective physical reality upon which the sensation depends is the length of the electromagnetic waves in the light beam. In various units these wavelengths are:

Color	Wavelength		
	Inches	Millimeters	Angstrom Units = A
Violet	0.000016	0.00040	4000
Blue	0.000018	0.00045	4500
Green	0.000021	0.00054	5400
Yellow	0.000023	0.00059	5900
Red	0.000026	0.00065	6500

An inch equals 25.4 millimeters; a millimeter equals 10,000,000 angstrom units.

By spreading the spectrum out into a long band the physicist can subdivide each color into thousands of steps each far too small for the eyes to distinguish. Thus wavelength is just a precise way of denoting color. In the language of physics, a numerical "wavelength" replaces the less exact term "color." Wavelengths can also be used to describe vibrations similar to those of light but "colorless"—either too short (ultraviolet) or too long (infrared) for the eye to see.

In most spectroscopes the light is admitted through a narrow slit and the spectrum appears as a rainbow-colored ribbon formed by a continuous sequence of narrow images of this slit. Any particular image that stands out from the background is called a spectrum line. The position of a line may be specified with great accuracy by giving the wavelength to which it corresponds.

26 *Space Chemistry*

Bright lines come from a body of gas hot enough to give out light. Each chemical element gives a set of lines arranged in a characteristic pattern. If white light shines through a cool gas, *dark* lines in the same pattern will be gouged out of the continuous spectrum. A particular fixed pattern of bright or dark lines thus reveals the chemical nature of the source. In Figure 7, for example, the trained spectroscopist would recognize that the vertical lines of the comparison spectrum above and below the two strips of the stellar spectrum are due to iron. If he were in doubt, a few measurements would reassure him. He would also recognize the remarkable series of hydrogen lines in the spectrum of the star.

Spectrum lines serve not only to identify the atom by which they were produced, but also to tell a good deal about the circumstances of the gas in which they originated. The temperature, for example, can be determined from the relative intensities of certain key lines. Should the temperature be high enough to ionize the atom, that is, to remove one of the outer electrons, the spectrum then

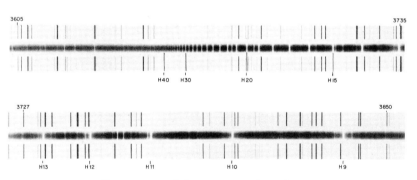

FIG. 7. Two sections of the spectrum of the faint star HD 193182 with comparison lines from an iron arc above and below. The star spectrum is crossed with absorption lines whose origin can be readily recognized by their positions. In this star the strong hydrogen lines H9 to H40 make a striking pattern.

undergoes a radical change. The atom is then represented by a completely different pattern of lines, for positions of the lines depend on the number and arrangement of the outer electrons. Molecules can be distinguished because their spectra are totally different from those of the atoms of which they are composed. The spectrum of the titanium oxide molecule TiO, for example, has a line pattern totally unlike that formed by the titanium atom or the oxygen atom. A great variety of chemical and physical data can thus be obtained from a detailed study of spectrum lines.

Sir William Huggins (1824–1910), an English amateur astronomer, was the great pioneer of cosmic chemistry. The story of his work is best told in his own words which we quote from *An Atlas of Representative Stellar Spectra,* published in 1899.

"The Observatory was erected in connection with my private dwelling-house at 90, Upper Tulse Hill, London, in 1856." [Its principal instrument was a telescope with an 8-inch lens made by Alvan Clark.]

"I soon became a little dissatisfied with the routine character of ordinary astronomical work, and in a vague way sought about in my mind for the possibility of research upon the heavens in a new direction, or by new methods. It was just at this time, when a vague longing after newer methods of observation for attacking many of the problems of the heavenly bodies filled my mind, that the news reached me of Kirchhoff's great discovery of the true nature and the chemical constitution of the sun from his interpretation of the Fraunhofer lines.

"Here at last presented itself the very order of work for which in an indefinite way I was looking—namely, to extend his novel methods of research upon the sun to the other heavenly bodies. This was especially work for which I was to a great extent prepared, from being already familiar with the chief methods of chemical and physical research.

28   *Space Chemistry*

"Then it was that an astronomical observatory began, for the first time, to take on the appearance of a laboratory (Fig. 8). Primary batteries, giving forth noxious gases, were arranged outside one of the windows; a large induction coil stood mounted on a stand on wheels, so as to follow the positions of the eye-end of the telescope, together with a battery of several Leyden jars; shelves with Bunsen

FIG. 8. The 8-inch telescope with which Sir William Huggins did his pioneer work a hundred years ago on the chemistry of the stars. He described it: "The illustration shows the interior of the observatory as it was in use for spectroscopic work from 1860 to 1869. The Spectroscope is attached to the eye-end of the 8-inch refractor with which the earliest work on the spectra of stars was done. The induction coil and the wires therefrom for the production of chemical spectra for direct comparison in the spectroscope with the celestial spectra are shown."

burners, vacuum tubes, and bottles of chemicals, especially of specimens of pure metals, lined its walls.

"The observatory became a meeting-place where terrestrial chemistry was brought into direct touch with celestial chemistry. The characteristic light-rays from earthly hydrogen shone side by side with the corresponding radiations from starry hydrogen, or else fell upon the dark lines due to the absorption of hydrogen in Sirius or in Vega. Iron from our mines was line-matched, light for dark, with stellar iron from opposite parts of the celestial sphere. Sodium, which upon the earth is always present with us, was found to be widely diffused through the celestial spaces.

"The time was, indeed, one of strained expectation and of scientific exaltation for the astronomer, almost without parallel; for nearly every observation revealed a new fact, and almost every night's work was red-lettered by some discovery."

Huggins' first observations of star spectra were visual, but, as he relates, "in February 1863 the strictly astronomical character of the Observatory was further encroached upon by the erection, in one corner, of a small photographic tent, furnished with baths and other appliances for the wet collodion process. We obtained photographs, indeed, of the spectra of Sirius and Capella; but from want of steadiness and more perfect adjustment of the instruments, the spectra, though defined at the edges, did not show the dark lines, as we expected. The dry collodion plates then available were not rapid enough; and the wet process was so inconvenient for long exposures, from irregular drying and draining back from the positions in which the plates had often to be put, that we did not persevere in our attempts to photograph the stellar spectra. I resumed them with success in 1875, as we shall see further on."

Dr. Henry Draper, an American professor of physiology and chemistry at the University of the City of New York, was the first to photograph lines in the spectrum of a star. In August 1872, using gelatine dry plates, he photographed the spectrum of Vega with the 11-inch refractor in his private observatory near Dobbs Ferry, New York. The plates showed four dark lines, indicating the presence of hydrogen. Draper later obtained photographs of the spectra of other bright stars and of the Orion nebula.

When improved photographic plates became available, Huggins resumed his attempts to record stellar spectra, and this time met with great success. In 1876 he recorded spectra of Vega showing seven strong dark lines, and from then on progress was rapid. To quote again from his account: "In the spectra of such stars as Sirius and Vega, there came out in the ultra-violet region, which up to that time had remained unexplored, the completion of a grand rhythmical group of strong dark lines, of which the well known hydrogen lines in the visible region form the lower members. Terrestrial chemistry became enriched with a more complete knowledge of the spectrum of hydrogen from the stars. Shortly afterwards, Cornu succeeded in photographing a similar spectrum in his laboratory from earthly hydrogen."

There is striking diversity in the patterns of dark lines which different stars exhibit in their spectra. The first to make a systematic classification of stellar spectra based on the patterns of these dark lines was the great Jesuit astronomer Father Angelo Secchi (1818–78) of the Collegio Romano Observatory. In 1863, after visually examining about 2000 stars, Secchi noticed a correlation between the colors of the stars and the dark lines in their spectra. The spectra of *white stars,* Type I, have, he reported, "an almost uniform prismatic series of colours, interrupted only by four very strong black lines." These are the great hydrogen lines which have since been so intensively studied. They are widely spaced over the visual spectrum, the first being in the red, others in the blue-green, blue, and violet (Fig. 9). *Yellow stars,* Type II, Secchi found to "have a

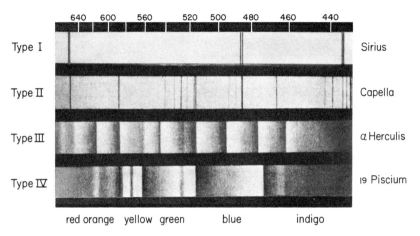

FIG. 9. Secchi's four types of stellar spectra from visual observations. Type I includes Harvard classes B, A, F0; Type II, F5, G, K; Type III, M; Type IV, N.

spectrum exactly like that of our sun—that is, distinguished by very fine and numerous lines." Nearly all these lines, we now know, are due to the presence of metals, such as sodium, magnesium, calcium, titanium, vanadium, and iron. *Red stars*, Types III and IV, had quite different spectra with wide dark spaces caused by molecules, rather than by single atoms. White stars are the hottest and red the coolest, with yellow intermediate. Many decades passed before the relation between the colors or temperatures of the stars and their apparent chemical composition as revealed by the spectrum lines was understood.

Secchi's survey was visual, but photography soon made spectroscopic surveys of the stars much easier and quicker. A great advance in our general knowledge of stellar chemistry was made about 1890 when a group of observers at Harvard, led by E. C. Pickering, made an extensive survey of star spectra as photographed from Cambridge, Massachusetts, and, for the southern sky, from Arequipa, Peru.

These observers worked out a new scheme of classification which resembled Secchi's but was more detailed and gave more chemical information. Their system proved so useful that it was soon adopted universally. Miss Annie J. Cannon, a member of the Harvard staff, recorded and later published the classifications of about 360,000 stars. This vast catalogue of eleven volumes is now a standard reference work. Some of the more general results will be described briefly in Chapter VI.

Our knowledge of the elements in the sun was increased by the work of Henry A. Rowland, professor of physics at Johns Hopkins University. Since all the fine detail in the spectrum of the sun can be brought out only by large-scale photographs at excellent definition, Rowland found it best to use diffraction gratings consisting of polished metal surfaces on which a large number of parallel lines were ruled very close together. On the resulting photographs of the sun's spectrum, Rowland and his collaborator L. E. Jewell measured about 23,000 individual spectrum lines. Chemical identifications were made by comparing these with the spectrum lines of pure elements, which Rowland also photographed and listed with high accuracy.* Identification of the lines has been extended and improved by scientists working with new photographs taken at the Mount Wilson Observatory.

For many decades after the door to celestial chemistry was opened by Kirchhoff and Bunsen, astrophysics was concerned with identifying chemical elements in the sun and stars—with what a laboratory chemist would call *qualitative* analysis. It was soon proved that many of the more abundant terrestrial elements were present in heavenly bodies, but for a long time it seemed that the heavens might contain a few chemical substances not found on earth. A narrow ray of green light, emitted by gaseous

---

*His results are recorded in an elaborate *Photographic Map of the Normal Solar Spectrum* (1888) and in the enormous *Preliminary Table of Solar Spectrum Wave-Lengths* (1898).

nebulae not matched by any light produced in the laboratory, was therefore assigned to a new element, nebulium. The intriguing stories of nebulium, and of another new element helium, eventually found on earth, are told in the following chapter.

Since 1929 emphasis has shifted to *quantitative* chemical analysis. In that year Henry Norris Russell, professor of astronomy at Princeton University, made such an analysis of the sun's atmosphere. Many similar investigations have since been made on other stars. As the results become more detailed and more reliable, astronomers are using them to study the origin as well as the physical and chemical development of stars and galaxies.

III. INDIVIDUAL CHEMICAL ELEMENTS

The total number of individual atoms in the observable universe is, mathematically, not infinite; but it is unthinkably large. A rough estimate is 10^{78}, or 1 with 78 zeros after it. The number of different kinds of atoms, on the other hand, is small enough to comprehend more easily. Only 103 elements are now known and all of them fall naturally into a single sequence according to the electric charge on the atomic nucleus.

Every atom has a stable inner core called its nucleus which always carries a positive electric charge. Nuclei are built up of two kinds of elementary particles, each of which is a basic unit. First of these is the *proton*, which always carries one unit of positive electric charge. The total charge on the nucleus is equal to the number of protons it contains, one for hydrogen, two for helium, three for lithium, and so on up to 103 for lawrencium. The whole array of chemical elements can be arranged as a series of successive integers called *atomic numbers.*

The second kind of elementary particle inside the nucleus is the *neutron*, which has no electric charge at all, but nevertheless performs a most important function. Every atomic nucleus except that of hydrogen contains two or more protons all of which have positive electric charges and therefore repel each other strongly when brought close together. The neutrons are somehow able to keep the protons from flying away from each other, and to bind them together inside the very small nucleus. The neutrons do this without in any way neutralizing the charges; the protons still exert their normal electric forces on all outside electrified particles. Just how neutrons manage this essential service is not yet clearly understood. A neutron outside an atomic nucleus is unstable; left to itself it will separate in about fifteen minutes into a positive proton and a negative electron.

In every atom, surrounding the stable nucleus, are a number of *electrons,* each with one unit of negative electric charge. The size of the unit is the same as that of the unit of positive charge on a proton. Each atom normally has as many outer electrons as there are protons in the nucleus. Thus the charges balance and the atom as a whole is electrically neutral. In a gas at high temperature, collisions of atoms are frequent and violent, and some of the electrons may be knocked off. An atom that has lost one or more electrons is no longer electrically neutral and is said to be ionized. These incomplete atoms give a pattern of spectrum lines totally different from that of normal atoms and may easily be recognized on spectroscopic photographs. In spectra of many stars, both normal and ionized atoms of the same element are recorded. The relative strengths of the two groups of lines depend on the temperature and pressure of the outer stellar layers of gas where the lines originate.

The *atomic weight* of an atom is very nearly the sum of the weights of the protons and neutrons in the nucleus. The electrons, each of which weighs only about 1/2000 as much as a proton, add very little. Protons and neutrons have about the same weight. Because most nuclei have a few more neutrons than protons, the atomic weight of an atom is usually somewhat more than twice the atomic number.

The first few types of atoms, beginning with hydrogen and helium, are often referred to as the light elements. The heavier elements include most metals.

TABLE 1

List of the Chemical Elements

Element	Symbol	Atomic Number	Approx. Atomic Weight	Relative Number of Atoms*
Hydrogen	H	1	1	25,000,000.
Helium	He	2	4	3,800,000.
Lithium	Li	3	7	0.1
Beryllium	Be	4	9	0.02
Boron	B	5	11	0.02
Carbon	C	6	12	9300.
Nitrogen	N	7	14	2400.
Oxygen	O	8	16	25000.
Fluorine	F	9	19	1.6
Neon	Ne	10	20	14000.
Sodium	Na	11	23	44.
Magnesium	Mg	12	24	910.
Aluminum	Al	13	27	95.
Silicon	Si	14	28	1000.
Phosphorus	P	15	31	10.
Sulfur	S	16	32	380.
Chlorine	Cl	17	35	2.6
Argon	A	18	40	150.
Potassium	K	19	39	3.2
Calcium	Ca	20	40	49.
Scandium	Sc	21	45	.03
Titanium	Ti	22	48	1.7
Vanadium	V	23	51	.2
Chromium	Cr	24	52	7.8

CONTINUED

Individual Chemical Elements 37

List of Chemical Elements CONTINUED

Element	Symbol	Atomic Number	Approx. Atomic Weight	Relative Number of Atoms*
Manganese	Mn	25	55	6.8
Iron	Fe	26	56	150.
Cobalt	Co	27	59	1.8
Nickel	Ni	28	59	27.
Copper	Cu	29	64	.2
Zinc	Zn	30	65	.2
Gallium	Ga	31	70	.009
Germanium	Ge	32	73	.03
Arsenic	As	33	75	.002
Selenium	Se	34	79	.02
Bromine	Br	35	80	.004
Krypton	Kr	36	84	.04
Rubidium	Rb	37	85	.006
Strontium	Sr	38	88	.06
Yttrium	Y	39	89	.009
Zirconium	Zr	40	91	.014
Niobium	Nb	41	93	.0008
Molybdenum	Mo	42	96	.002
Technetium	Tc	43	99	—
Ruthenium	Ru	44	102	.0009
Rhodium	Rh	45	103	.0002
Palladium	Pd	46	107	.0007
Silver	Ag	47	108	.0003
Cadmium	Cd	48	112	.0009
Indium	In	49	115	.0001
Tin	Sn	50	119	.001
Antimony	Sb	51	122	.0002
Tellurium	Te	52	128	.003
Iodine	I	53	127	.0006
Xenon	Xe	54	131	.003
Cesium	Cs	55	133	.0005
Barium	Ba	56	137	.004
Lanthanum	La	57	139	.0005
Cerium	Ce	58	140	.0006
Praseodymium	Pr	59	141	.0002

CONTINUED

List of Chemical Elements CONTINUED

Element	Symbol	Atomic Number	Approx. Atomic Weight	Relative Number of Atoms*
Neodymium	Nd	60	144	.0009
Promethium	Pr	61	147	—
Samarium	Sm	62	150	.0002
Europium	Eu	63	152	.0001
Gadolinium	Gd	64	157	.0005
Terbium	Tb	65	159	.00009
Dysprosium	Dy	66	162	.0007
Holmium	Ho	67	165	.0002
Erbium	Er	68	167	.0006
Thulium	Tm	69	169	.00009
Ytterbium	Yb	70	173	.0004
Lutetium	Lu	71	175	.00004
Hafnium	Hf	72	179	.0001
Tantalum	Ta	73	181	.00002
Tungsten	W	74	184	.0001
Rhenium	Re	75	186	.00005
Osmium	Os	76	190	.0006
Iridium	Ir	77	193	.0005
Platinum	Pt	78	195	.001
Gold	Au	79	197	.0001
Mercury	Hg	80	201	.0004
Thallium	Tl	81	204	.0003
Lead	Pb	82	207	.02
Bismuth	Bi	83	209	.0003
Polonium	Po	84	210	—
Astatine	At	85	211	—
Radon	Rn	86	222	—
Francium	Fr	87	223	—
Radium	Ra	88	226	—
Actinium	Ac	89	227	—
Thorium	Th	90	232	.00003
Protactinium	Pa	91	231	—
Uranium	U	92	238	.000008
Neptunium	Np	93	237	—
Plutonium	Pu	94	242	—
Americium	Am	95	243	—

CONTINUED

Individual Chemical Elements

List of Chemical Elements CONTINUED

Element	Symbol	Atomic Number	Approx. Atomic Weight	Relative Number of Atoms*
Curium	Cm	96	243	—
Berkelium	Bk	97	245	—
Californium	Cf	98	246	—
Einsteinium	E	99	253	—
Fermium	Fm	100	254	—
Mendelevium	Mv	101	254	—
Nobelium	No	102	—	—
Lawrencium	Lw	103	—	—

*The number of atoms or cosmic abundance in the fifth column are relative to a value of 1000 for silicon, atomic number 14. They are from A. G. W. Cameron's revision of an earlier list by H. E. Suess and H. C. Urey.

The abundance of light elements in the universe has been determined by spectroscopic observations of the sun, stars, and nebulae; that of heavier elements largely by laboratory chemical analyses of meteorites. Table 1 lists, in addition to the atomic numbers of elements, their relative abundance. Silicon, of medium abundance, is the standard of reference with an arbitrary number of 1000. All other numbers are relative to this one.

The really abundant elements are hydrogen (atomic number 1) and helium (2). Hydrogen has 25,000 times as many atoms as silicon, helium 3800 times as many. Next in order of abundance are oxygen (8), neon (10), carbon (6), and nitrogen (7). Magnesium (12) is almost as abundant as silicon (14), and is followed by iron (26), aluminum (13), and calcium (20), all of which are of great astronomical interest. Other notable elements are sulfur (16) and argon (18). Nickel (28) has about one-fifth as many atoms as iron (26). For heavier metals the proportions are extremely small; nevertheless, strontium (38), yttrium (39), zirconium (40), and barium (56) are of considerable astronomical interest as are also the elements known to chemists as the "rare earths" from lanthanum (57) to lutetium (71).

In Table 1 the chemical elements are arranged in order of atomic number. A better arrangement is the well-known Periodic Table discovered by the Russian chemist Dmitri Mendeléieff about 1870. According to this table all elements may be placed in nine groups (columns 0 to 8 of Table 2), the elements in each group having similar properties. Thus in Column 0 the elements helium, neon, argon, krypton, and xenon are all inert gases, none of which enters readily into compounds of any kind. Within the other columns the elements have similar properties and generally form the same types of compounds. It is now known that these properties depend on the arrangement of the electrons, particularly on the number of electrons in the outermost "shell."

In each horizontal period of Table 2 the elements are listed in the order of number of outer electrons which are left over after a complete shell of electrons has been formed. These are the "valence" electrons which take part in the combinations of atoms called molecules, and are responsible for most of the chemical and spectroscopic properties of each kind of atom. In each vertical group all elements have the same number of outer or valence electrons. Group 0 of Table 2 (listed separately with certain other data in Table 3) marks the completion of an electron shell and Group 1 marks the first step in building up a new shell. The pattern of the characteristic spectrum lines of any element depends on these same outer electrons. Hence in spectroscopic studies the elements present the same groupings that they do in their purely chemical properties. For this reason the periodic grouping in Table 2 is of much interest to astronomers as well as to laboratory chemists. The spectrum lines from corresponding elements in a group such as lithium, sodium, and potassium form

TABLE 2
Periodic Arrangement of the Chemical Elements

Group Period	0	I	2	3	4	5	6	7	8
I		1. H							
II	2. He	3. Li	4. Be	5. B	6. C	7. N	8. O	9. F	
III	10. Ne	11. Na	12. Mg	13. Al	14. Si	15. P	16. S	17. Cl	
IV	18. A	19. K	20. Ca	21. Sc	22. Ti	23. V	24. Cr	25. Mn	26. Fe 27. Co 28. Ni
		29. Cu	30. Zn	31. Ga	32. Ge	33. As	34. Se	35. Br	
V	36. Kr	37. Rb	38. Sr	39. Y	40. Zr	41. Nb	42. Mo	43. Tc	44. Ru 45. Rh 46. Pd
		47. Ag	48. Cd	49. In	50. Sn	51. Sb	52. Te	53. I	
VI	54. Xe	55. Cs	56. Ba	57–71	72. Hf	73. Ta	74. W	75. Re	76. Os 77. Ir 78. Pt
		79. Au	80. Hg	81. Tl	82. Pb	83. Bi	84. Po	85. At	

similar but not identical patterns. Each element can be distinguished without ambiguity.

Now we are ready to study a few of the individual elements most important in cosmic chemistry.

Hydrogen

Hydrogen takes its name from the fact that when it is burned in air, water is produced. It is the simplest, the lightest, and the most abundant element in the universe. The spectrum lines of hydrogen are strong and many of them are in the blue and violet where they are easy to photograph. These lines have probably been observed in more heavenly bodies than those of any other element.

TABLE 3
Inert Gases

Atom		Nucleus		Atomic Weight	Electrons
Name	Symbol	No. Protons =Atomic No.	No. Neutrons		
Helium	He	2	2	4	2
Neon	Ne	10	10	20	10
Argon	A	18	22	40	18
Krypton	Kr	36	48	84	36
Xenon	Xe	54	77	131	54
Radon	Rn	86	136	222	86

Hydrogen was probably known to medieval chemists, but scientific knowledge of it may be said to have begun in 1766 with the work of Henry Cavendish in London. Hydrogen lines were first seen in the solar spectrum by W. H. Wollaston in 1802, and were studied more closely both in the sun and stars by J. Fraunhofer in 1814. Their chemical significance was not recognized, however, for many years. They apparently were not observed in the laboratory until 1851 when A. Masson of Paris noticed them in the light of electric sparks seen through a prism. The hydrogen came from water vapor in the air. In 1862, a few years after the general principles of spectrum analysis had been made clear by Kirchhoff and Bunsen, A. J. Angström of Uppsala announced the presence of hydrogen and of ten metals in the solar atmosphere. In 1876 Sir William Huggins photographed numerous hydrogen lines in the ultraviolet part of the spectrum of the bright star Vega. At that time, only a few of the stronger of these lines had been observed in the laboratory. Ever since then the stars have kept ahead of the laboratory in the number of observed hydrogen lines.

Atoms are often said to be like miniature solar systems. The heavy nucleus corresponds to the sun, the lighter electrons to the encircling planets. The electric force which the positively charged nucleus exerts on the negative electron takes the place of the gravitational attraction of the sun on a planet. In the hydrogen atom there is just one electron, and it is easy to understand how this might be in a stable orbit which repeats itself indefinitely just as the earth can go around the sun in the same orbit year after year. But what happens when the atom emits or absorbs a tiny bit of light in one of the atom's characteristic colors? Light carries energy and the loss or gain of energy would disturb the electron in its orbit. What would happen, and how the electron would ever get back to its normal orbit were theoretical problems with which physicists struggled for many years without success. Finally in 1913, Niels Bohr of Copenhagen suggested a set of postulates, now called "quantum theory," which gave a satisfactory solution to the problem, and began a whole new chapter in our understanding of the structure of atoms. Bohr's theory of the hydrogen atom was based not on one orbit for the electron, but on a whole series of possible orbits of different sizes. When the electron jumped outward from one orbit to a larger orbit, it *absorbed* a bit of

44   *Space Chemistry*

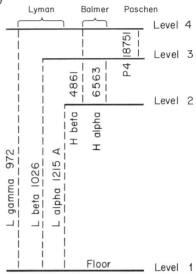

FIG. 10. Energy levels of the hydrogen atom and their relation to the spectral series. Horizontal lines show the only energy levels the outer electron can have. The vertical lines with wavelengths indicate the lines formed when the electron jumps from one level to another.

light of a definite wavelength (a spectrum line) and when it jumped inward to a smaller orbit it *emitted* a line. Some rather simple but novel theoretical assumptions accounted for the observed wavelengths of the lines.

The real spirit of Bohr's hypothesis can perhaps be conveyed by a model, a figure of speech if you like, totally different from the notion of orbits. Imagine a hydrogen atom to resemble a staircase. The electron is a rubber ball normally lying on the floor at the foot of the staircase. If hit by a stick it bounces up onto one of the steps at a definite height above the floor. These are the only places where it can go; it cannot stay in midair. Then after a short time it can jump down again to any lower step or back to the floor. Every jump corresponds to a light ray, a spectrum line. A jump up corresponds to absorbing a line, a jump down to emitting one. The longer the jump, the shorter is the wavelength of the line. The successive steps of the staircase represent what the physicist would call discrete energy levels which the electron can occupy.

The steps of the imaginary staircase are not of equal height but decrease regularly in height from the floor up. The relative heights are shown in Figure 10, where the horizontal lines represent the floor and the successive steps in the staircase. The vertical dotted lines represent the various possible jumps, each one corresponding to a spectrum line. The longer the vertical line the more energy and the shorter the wavelength of the corresponding spectrum line. The details of this correspondence are for the specialist.

A simpler diagram of the various groups of hydrogen lines is shown in Figure 11. Each group, called a "series," is known by the name of the man who discovered it or investigated it most thoroughly. At the lower left is the Lyman series in the far ultraviolet where the air is quite opaque. It has been photographed in the laboratory with vacuum spectrographs, and recently in the spectrum of the sun from rockets flying high enough to be above most of the earth's atmosphere.

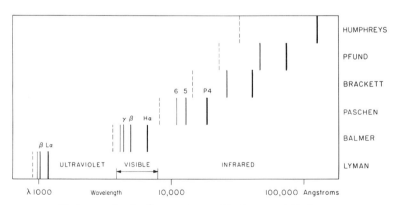

FIG. 11. Various series in the spectrum of hydrogen plotted on a logarithmic scale. The Balmer series is the only one in the visual region.

Next comes the Balmer series which thus far has yielded most of our knowledge about hydrogen in the stars. Five or six lines are visible to the eye and many others in the near ultraviolet are easily accessible through photography. A few lines in the infrared Paschen series have been photographed on special plates in the sun and stars.

In the spectra of most stars, the hydrogen lines are dark because the hydrogen atoms absorb more light than they emit. In some stars the emission is stronger than the absorption and the lines are bright. This may happen because the star is surrounded by a large cloud of glowing hydrogen, or for other reasons. Examples of bright and dark hydrogen lines are shown in Figure 12. The red Hα line in emission is easily recognized even in low dispersion spectra of faint stars. Hundreds have been found on red-sensitive plates that have been obtained by mounting an objective prism in front of the lens of the telescope (Fig. 13) or by slitless spectrographs. Many of the violet lines of hydrogen can be observed in the spectra of hot stars by this method.

In addition to the spectrum lines which can be seen and photographed, hydrogen atoms send out waves similar to light waves, but 21 centimeters long, which can neither be seen nor photographed. They must be observed by radio methods. These waves are of great interest to astronomers because they are freely emitted by the cool hydrogen atoms in interstellar space. They have already provided important information concerning the structure of the Galaxy that could not be obtained from the ordinary spectrum lines of hydrogen. Some of the results will be described in Chapter V.

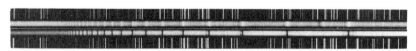

FIG. 12. The Balmer series in emission (above) in BD+11°4673 and in absorption in 48 Librae (below). The intensity and separation of the lines increase toward the right (longer wavelengths). (Positive.)

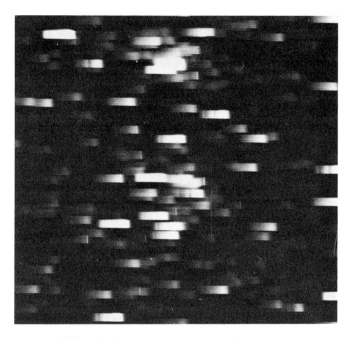

FIG. 13. Widened objective-prism spectra of stars in the double cluster in Perseus with the 10-inch refractor and prism in front of the lens. For the brighter stars the region from D_1D_2 (sodium) on the left to hydrogen Hα on the right appears. In several stars Hα in emission is easily seen. Some spectra show molecular bands.

Helium

Helium was discovered on the sun long before it was recognized in the laboratory. Next to hydrogen it is the most abundant element in the universe. In stars and nebulae there are nearly one-sixth as many helium atoms as hydrogen atoms, while oxygen ranks a poor third with only 1/1000 as many. In B-type stars the helium lines are the most prominent lines of their spectra (Figs. 14 and 15).

Its recorded history began in 1868 during an eclipse of the sun, when the French astronomer P. J. C. Jansen saw a bright yellow line not ordinarily visible in the sun's spectrum. It came from a prominence on the sun's limb. This

48   Space Chemistry

FIG. 14. Helium lines in χ Aurigae.

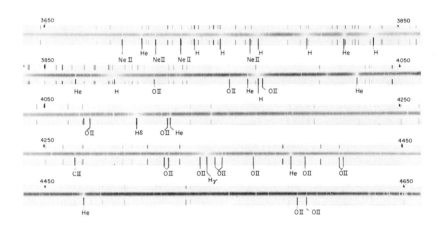

FIG. 15. Lines of helium (HeI), carbon (CII, 4267A), oxygen (OII), and neon (NeII) in the hot star τ Scorpii.

line was not the same as any previously observed, and the English astronomer Sir Norman Lockyer, who also observed it, concluded that it originated in a gas as yet unknown on earth. For this hypothetical element he coined the name "helium" after the Greek word for sun.

Twenty-seven years later, in 1895, Sir William Ramsay found a laboratory gas which produced the same yellow line discovered in the sun. This gas was obtained by treating with acid the minerals uraninite and cleveite. At first it was called cleveite gas but before long the name helium was generally adopted.

On earth helium is far less abundant compared to other elements than in stars and nebulae. Small amounts have been found in rocks of the earth's crust and in natural gas trapped underground. In certain areas in Texas, Oklahoma, and Kansas the proportion is unusually large, ranging from 1 to 8 per cent, and nearly all the helium in the western world comes from here. It is extracted by a freezing process which liquefies all other gases. (Helium has the lowest boiling point of all gases.) The present rate of production is about 635,000,000 cubic feet per year, but the demand is growing and will probably exceed the supply in a very few years.

Helium is used to fill balloons. It is indispensable in low temperature research and is helpful in experimental wind tunnels where very high velocities are to be studied. Because it is chemically inert and a good conductor of heat, it is used in arc-welding of various metals. It is used also in the production of titanium and zirconium and in the preparation of silicon and germanium crystals for electrical transistors. In the future it may be needed for gas-cooled reactors in atomic energy plants.

There is plenty of helium in the air, nearly 200,000 pounds above every square mile of the earth's surface, but its fractional part of the whole atmosphere is so small, only about 0.0005 of 1 per cent, that the expense of extracting it would probably be prohibitive. The mantle of Jupiter, well below the visible surface of the planet, may contain much liquid or solid helium; but many years will elapse before we can tap this source of supply.

Carbon

The chemical elements known to the ancient world included six or seven metals and two nonmetals, carbon and sulfur. Charcoal, which is fairly pure carbon, was prepared at least as early as the first century A.D. by heating wood covered with clay to exclude air.

Carbon is an outstanding element in astrophysics. According to recent tabulations (see Table 1), only hydrogen, helium, oxygen, and neon surpass carbon in the cosmic abundance of their atoms.

Atomic carbon in the sun and stars is not easy to study because its chief spectrum lines are in the far ultraviolet and cannot be photographed from the surface of the earth. Except for the 4267 A line of ionized carbon which is prominent in hot stars of types O and B (Fig. 15), we must be satisfied to observe secondary and relatively weak lines.

Certain minor lines of the carbon atom which lie in the infrared are fairly strong in the solar spectrum. The actual number of carbon atoms in the sun's atmosphere must be quite large, for a spectrum line increases in strength with the amount of the element which produces it.

In the interiors of all stars and in the outer layers of most of them, the temperature is too high for carbon compounds to exist. Thus most of the carbon in the universe is in atomic form. Molecules containing carbon do exist, however, in certain relatively cool places, such as atmospheres of planets and cooler stars, comets, and interstellar space.

Oxygen

Oxygen, possibly the most important element on earth, was discovered in the 1770's independently by Joseph Priestley in England and by C. W. Scheele in Sweden. The name "oxygen" was provided by the French scientist A. L. Lavoisier, who is said to have been the first to assert that the substance is a chemical element. He coined the name from the Greek word for sharp because he mistakenly believed the element to be an essential constituent of acids.

The spectrum lines of oxygen were first recognized by A. J. Ångström in 1853, and the wavelengths of several were measured by J. Plücker a few years later. A valid identification of oxygen lines in astronomical spectra was made about 1897 by the Englishman F. McLean in several high-temperature stars of the southern sky. He mentioned especially Beta Crucis, which he called an "oxygen star" because of the number and prominence of dark lines of oxygen in its spectrum. Since that time other astronomical sources have disclosed many lines produced by oxygen atoms under various physical and chemical circumstances (Fig. 15), and oxygen is now quite important in astrophysics. Its atomic lines have been extensively studied in many stars, especially the hotter ones. In gaseous nebulae like the famous Orion nebula, the green color is due to oxygen atoms [OIII] which have lost two of their outer electrons (Figs. 60 and 62). In the coolest stars, oxygen is recognizable in compounds with metals, because these compounds, called oxides, produce striking, unsymmetrical features (molecular bands) in stellar spectra. The strongest bands of this kind are those of titanium oxide and zirconium oxide.

Neon

Neon was discovered in 1898 by Sir William Ramsay and Dr. Morris W. Traverse by liquefying argon obtained from the earth's atmosphere and then allowing it to warm up slowly. The first gas to come out of the liquid was an impurity not previously known. When put in an electrical discharge tube it gave out a "magnificent" spectrum with many lines in the yellow and red. The discoverers called it "neon," after the Greek word for new.

Although rare on earth, neon is abundant in stars and nebulae. Dark spectrum lines corresponding exactly in wavelength to the bright lines coming from a neon sign testify to the presence of neon in many of the hotter stars. The visible star Upsilon Sagittarii, one of the most peculiar objects in the sky, is noted for the strength of the neon lines in its spectrum. Extremely hot stars have lines of ionized neon (Fig. 15).

In gaseous nebulae, including both large ones like the well-known Orion nebula and the smaller "planetary" nebulae, neon glows strongly—but not with its familar red color. In nebulae, conditions of incandescence are quite different from those in neon signs; much of the neon is ionized and glows with a violet tinge. Neon has been recognized with certainty in distant galaxies like the famous spiral in Andromeda.

Neon is one of a group of elements called inert gases because they are gaseous except at extremely low temperatures and because their atoms have no tendency to join with other atoms to form compounds. The inert gases have been called the "rare" gases but the adjective is inappropriate because helium, neon, and argon are, cosmically speaking, not rare at all but quite abundant.

In the earth's atmosphere, amounts of the inert gases occur in this order: argon, neon, helium, krypton, zenon. In the universe at large the distribution is very different; here helium is far ahead of the others, ranking next to hydrogen; and neon comes next, being five times as abundant as argon.

Metals

Thus far we have considered the abundant elements hydrogen, helium, carbon, oxygen, and neon. We now turn to certain metals whose atoms, while far less abundant, are nevertheless astronomically important. Lines of most metals are weak or absent in spectra of the hotter stars, but in passing along the temperature sequence toward cooler stars they gradually become dominant. In the spectrum of our sun, for example, metallic lines are far more numerous than those of the light elements previously discussed. They are also stronger, with the exception of two or three lines of hydrogen. In cooler stars the hydrogen lines are still weaker, and the dominance of metals becomes virtually complete as in the spectrum of Arcturus (Fig. 5).

Lithium, atomic number 3, is extremely scarce. Its behavior seems irregular and is not well understood. Further studies of lithium in star atmospheres may have an important bearing on theories of star structure and evolution. At high temperatures in the interiors of stars, lithium should disappear rapidly by combining with hydrogen to form helium. Thus the presence of lithium in the atmosphere of a star would appear to indicate that the general circulation of matter between atmosphere and interior must be very slow.

Sodium, atomic number 11, although not of great cosmic abundance, is widespread throughout the universe. Its distinctive spectrum lines, the famous yellow D lines, are well known in both laboratory and astronomical spectra. They are present in spectra of comets and all stars except the hottest, in whose atmospheres virtually all sodium atoms have lost one electron (thus giving a different set of spectrum lines), or some of the coolest where the lines may be hidden by molecular bands. The presence and motions of sodium atoms in the gas between the stars has been extensively studied (Fig. 16).

Magnesium, atomic number 12, is nearly twenty times as abundant as calcium, but its chief spectrum lines lie in the ultraviolet where until recently the earth's atmosphere has prevented their observation in astronomical sources. In the future, when spectra can be obtained from above most of the earth's atmosphere, these lines will rival and perhaps surpass in importance those of calcium, which are now so useful in astrophysics. Their strength in the sun's spectrum has already been proven by photographs taken from rockets (Figs. 77 and 78).

Calcium, atomic number 20, is of vast astrophysical importance, not because of its great abundance, but because

54 *Space Chemistry*

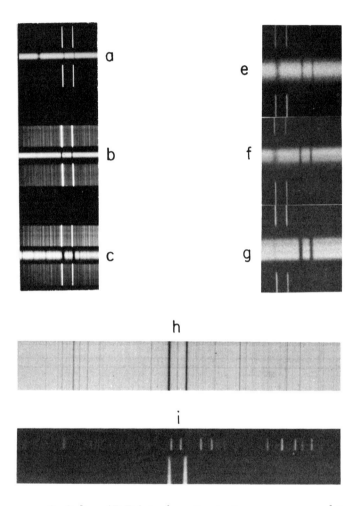

FIG. 16. Sodium (D_1D_2) in absorption in stars, space, sun, and in emission in comet. In different stellar types; a. β Ori B8, b. α Cyg. A2, c. α Ori M2: interstellar lines in spectra of hot stars at different distances; e. χ Aur (near), f. κ Cas (intermediate), g. 6 Cas (far), h. sun, and i. Comet Mrkos in emission.

it is a representative metal and because its chief spectrum lines lie in the violet part of the spectrum regularly photographed on ordinary plates.

In 1808 Sir Humphry Davy found that lime, long considered an elementary substance, is really a compound of a metal with oxygen. The metal he named "calcium" after the Latin word *calx,* meaning lime or chalk. In 1814 Fraunhofer gave to two strong dark lines in the solar spectrum (wavelengths 3968 A and 3933 A) the designations H and K by which they have been known ever since. In 1860 Kirchhoff and Bunsen identified these lines and an additional one at 4227 A with calcium. Later laboratory analysis, following the ground-breaking work of Sir Norman Lockyer, showed 4227 A to be due to the normal, electrically neutral atom with all its electrons, and H and K to the singly ionized atom (one electron removed).

These three lines of calcium are visible as dark lines on low-dispersion spectrograms of many stars (Fig. 17). The relative intensity of H and K with respect to 4227 A is a most useful measure of surface temperature. A few of the very hottest stars do not show any calcium lines at all; some not quite so hot show the H and K lines only; but the majority have 4227 A as well as H and K with 4227 A becoming relatively stronger as the temperature decreases. Moreover, in most star spectra calcium lines are stronger than lines of any other element: they are therefore

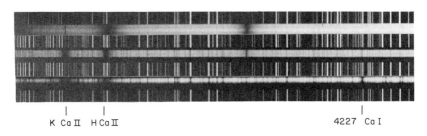

FIG. 17. Calcium in stars; Sirius A2 (top), Procyon F5, μ UMa K5.

56   Space Chemistry

important criteria for temperature classification throughout the whole range of stars.

Iron, atomic number 26, was probably known to the Egyptians as long ago as 2500 B.C. By the time of Christ it was in common use. Its introduction into warfare caused concern similar to that recently aroused by the use of the atomic bomb. Pliny the Elder, A.D. 23–79, wrote: "It is by the aid of iron, that we construct houses, cleave rocks, and perform so many other useful offices of life. But it is with iron also that wars, murders, and robberies are effected . . . For, as if to bring death upon man with still greater rapidity, we have given wings to iron and taught it to fly."

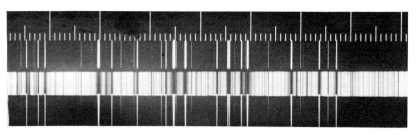

FIG. 18. Iron lines in the sun may be identified by the lines in the comparison iron arc.

FIG. 19. Absorption lines of iron in R Leonis may be identified by the iron lines in the comparison arc.

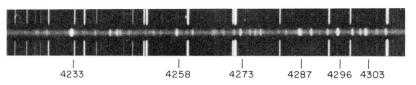

FIG. 20. Emission lines of iron (FeII) in XX Ophiuchi.

Iron is one of the three most important elements in the spectroscopic study of the stars (Figs. 18, 19, and 20), the other two being hydrogen and calcium. It is also the dominant constituent in meteorites—the only cosmic bodies of which samples can be examined in the laboratory for chemical and isotopic composition. Mixed with nickel and with smaller amounts of cobalt and copper, it forms from 70 to 94 per cent of the so-called "iron" meteorites; it is also abundant in "stony" meteorites.

Technetium

Technetium is of special interest in a chemical survey of the universe because, at present, it is the only element definitely identified in a heavenly body that has not been found in the crust or the atmosphere of the earth.

This element has a curious history. When in 1869 Dmitri Mendeléieff drew up the "Periodic Table" of the chemical elements then known, he found unfilled spaces which suggested the existence of elements related to boron, aluminum, silicon, manganese, and tantalum. We are interested here in the elements related to manganese, which he called "eka-manganese" and "dvi-manganese," the prefixes being the Sanskrit words for first and second. In 1925 three German physicists W. and I. Noddack and O. Berg announced the discovery of both elements. Eka-manganese, atomic number 43, which was identified only from lines in the X-ray spectrum, they called masurium. Dvi-manganese, atomic number 75, they called rhenium. Shortly thereafter W. F. Meggers of the U.S. Bureau of Standards requested from the Noddacks samples of both elements. In due course he received small quantities of rhenium from which he produced a spectrum whose lines were quite different from those of any other chemical element. But no "masurium" was ever received, and it now seems probable that the apparent identification was due to some misinterpretation of the laboratory data.

The first valid identification of element 43 was made years later in a totally different way. It was the first "artificial" element to be produced by the modern alchemy of the cyclotron. The preceding element in the chemical table, number 42, is molybdenum, a metal known to chemists since about 1780. If an additional proton could be forced into the nucleus of an atom of molybdenum, a nucleus of atomic number 43 would result. It occurred to an Italian physicist Emilio Segrè that the experiment might be worth trying. In 1936, visiting in Berkeley, California, he obtained samples of molybdenum which had been bombarded with the nuclei of deuterium (heavy hydrogen) atoms in the 37-inch cyclotron. (A deuterium atom has the same chemical properties as a hydrogen atom, but its nucleus is twice as heavy; instead of one proton alone as in ordinary hydrogen it has one proton and one neutron.) These samples he took back with him to Palermo, Italy, and in them the very next year with the assistance of a mineralogist Carlo Perrier he was able by chemical methods to identify element number 43, which he named technetium. It was mildly radioactive. In 1938, back in Berkeley he and Glenn T. Seaborg found a more stable form (isotope) of technetium. It is also slightly radioactive but has a half-life of about 200,000 years. Soon after, he and Miss Chien-Shiung Wu found technetium among the fission products of uranium. Later the Oak Ridge Laboratory of the Atomic Energy Commission produced appreciable amounts by a fission chain reaction, and by 1950 was able to supply samples of several milligrams. Thus Dr. W. F. Meggers of the U.S. Bureau of Standards eventually obtained a small amount of element number 43, now called technetium, which he caused to give out light by placing it in an electric arc. He then compiled an extensive and accurate list of the lines in its spectrum, which made it possible to search for technetium not only on earth but in the stars.

Chemists tried hard to find the new element in the crust of the earth. At one time it appeared that there might be traces of it in ores containing manganese and rhenium, but the consensus now is that there is no evidence of technetium in natural minerals. In the spectrum of the sun there are a few weak lines that might possibly be those of ionized technetium but the evidence is not convincing.

Atoms of technetium are not stable but are somewhat radioactive; half of them will disappear by spontaneous nuclear transformation in about 200,000 years. Thus it is not surprising that none now exists naturally on earth or in the sun. What is surprising is that moderate quantities have been found to exist in the atmospheres of S-type stars (Fig. 21). The explanation of the presence of an unstable element in these stars is not yet clear. Possibly they are very young and the initial supply of technetium is not yet exhausted. It may be that under certain circumstances a star can manufacture a little technetium as it goes along— perhaps in the interior where the temperature and pressure are tremendous, or, as a rather remote possibility, in the relatively cool outer atmosphere by the continual bombardment of cosmic rays. Nuclear physicists may soon be able to arrive at the correct explanation.

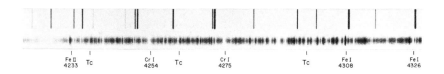

FIG. 21. Technetium (Tc) lines 4238 A, 4262 A, and 4297 A in the S-type variable star R Andromedae. Technetium is the one example of a chemical element found in the stars but not naturally present on the earth.

Californium

Californium like technetium is an artificial element, not known in natural minerals but produced in the laboratory by bombarding curium, element number 96, with helium nuclei, and in a thermonuclear bomb explosion by bombarding uranium, number 92, with neutrons. It is not stable but strongly radioactive; of a given number of atoms one-half will disintegrate spontaneously in 55 days. This has suggested a possible connection between californium and supernovae. A supernova is a rare, tremendously bright temporary star that appears without warning. The average rate of appearance in each galaxy is only one in about 500 years. The cause of the outburst is not well understood. One of the best observed supernovae, in galaxy IC 4182, was found, during the period of declining light after maximum, to drop to half brightness in 55 days, the same interval as the half-life of californium. This coincidence led to the suggestion that californium is created in a supernova outburst and somehow controls the rate of light emission while the star fades. Details are vague and the hypothesis is tentative.

Many of the lighter chemical elements are of especially great astronomical interest. I have therefore outlined the cosmic behavior of hydrogen, helium, carbon, oxygen, and neon. Calcium, iron, and a few other important members of the group of metallic elements were included. Technetium and californium were mentioned because they seem to be lacking on earth, except for infinitesimal manmade samples. It would not be profitable here to continue discussion of the elements through those of heavier atomic weight; some of them will be mentioned in Chapter V in connection with various astronomical bodies. The heavy elements are in general much scarcer than the light ones, and with some irregularities the abundance decreases toward the end of the list. We will therefore dismiss the remaining elements with the following brief comment.

Individual Chemical Elements

About two-thirds of all chemical elements found on earth, including all the more abundant ones, have been identified in astronomical objects. There is little reason to believe that the others are nonexistent. Astronomers may have failed to detect them for one or more of the following reasons: (1) the atoms are too scarce to produce detectable lines in celestial spectra; (2) the atoms have no outstanding lines in the parts of the spectrum that have yet been examined; (3) some heavy atoms might possibly lurk in the interiors of stars, never diffusing in detectable quantities up into the observable outer zones. Astronomers believe that the observable outer layers of a star are fairly good chemical samples of the unseen interiors, but these samples may possibly lack certain heavy elements.

IV. MOLECULES

The earth is a poor starting point for studies in space chemistry. It presents an unrepresentative chemical sample of the universe in two major ways: It has far less than the average proportions of hydrogen, helium, and neon; and its crust and atmosphere are not composed, like most of the mass of the universe, of separate atoms but instead are made up almost entirely of molecules.

The distinctive central nucleus of an atom is an extremely stable unit, held together by very strong forces. It is completely unharmed by the buffeting it receives in the rapid-fire collisions with other particles in a gas at temperatures of hundreds of thousands of degrees. Some of the outer electrons may be knocked off but the nucleus itself is strong enough to escape damage. Molecules are groups of two or more atoms held together by forces which are very feeble compared to the tremendous stabilizing forces which bind the protons and neutrons inside atomic nuclei. Most molecules break up into their individual atoms when in a gas heated to a few thousand degrees.

It is only in a few special places such as planets, comets, and in the outermost layers of certain stars where the gases are exceptionally cool that molecules can and do exist. On earth, molecules are all important.

Atoms produce spectrum lines which fall singly or in loose, nondescript, scattered groups. They are usually identified only by accurate measurements of their positions. Most molecules, on the other hand, produce closely packed sequences of lines usually forming unsymmetrical groups starting with a conspicuous sharp boundary and fading gradually toward either longer or shorter wavelengths. Scores of individual lines may be involved. The groups of lines are called bands, the sharp boundaries, band heads. Thus scattered individual lines are associated with atoms, band heads with molecules.

Cosmically speaking, most of the molecules we know about are in the atmospheres of the cooler stars where they exist in sufficient numbers to make their spectrum bands easy to observe. The types of molecules that exist in the atmospheres are those which best resist disruption by high temperature. They are chiefly in three groups: metal oxides, i.e., one atom of oxygen bound to one atom of a metal; carbon compounds; and compounds containing hydrogen.

Metal oxides. Because of the abundance of oxygen in stars, atoms of metals easily find oxygen atoms with which to combine to form oxides—when the temperature becomes low enough. The lower the temperature, the more numerous the molecules and the stronger the spectrum bands. In this matter, observation and chemical theory are in agreement. Molecules are few in the sun's disk but are frequent in the lower temperature of spots (Fig. 22).

Certain pronounced bands in stellar spectra, first sketched by Secchi and later photographed in great detail by many astronomers, were compared in 1904 by A. Fowler with his laboratory photographs of the spectrum of an electric arc into which titanium oxide was fed. The correspondence was so good that the existence of titanium oxide in stellar atmospheres was definitely proved. Other compounds have

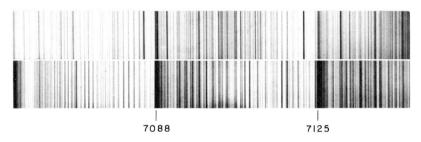

FIG. 22. Weak titanium oxide molecular bands at 7088 A and 7125 A in the spectrum of a sunspot (above) are identified by a comparison with a laboratory spectrum (below) in which the TiO bands are stronger.

been identified in a similar way in the long-period variable stars (Figs. 23 and 24). Oxides of the following metals are present in the outer layers of the cooler stars or in sunspots: boron, magnesium, aluminum, scandium, titanium, vanadium, chromium, manganese, strontium, yttrium, zirconium, niobium, and lanthanum.

Carbon compounds. In order to cause conspicuous markings in spectra of stars, atoms or molecules must be abundant and must have outstanding spectroscopic features in the parts of the spectrum most readily observable. Carbon molecules of two kinds fulfill both of these conditions and provide the characteristic features in the spectra of certain kinds of red stars. These features are more striking and have been known longer than the relatively weak lines of carbon atoms in hot stars.*

Carbon molecules of these two kinds are also well known on earth. They are molecules containing two atoms

---

*The band spectra of carbon molecules were among the first examples of emission spectra to be observed in the laboratory. In 1802 Wollaston saw five images of a candle flame viewed through a prism; these images, probably corresponded to emission bands of molecular carbon, C_2. The bands of cyanogen, CN, were seen in 1834 by Michael Faraday. A more extensive investigation of the spectra of flames was made in 1856 by W. Swan whose name is still attached to the chief bands of C_2. The line spectrum of atomic carbon was not observed until much later.

of carbon, C_2, and those of cyanogen, CN, containing one atom of carbon and one of nitrogen. Curiously these same molecules are the chief constituents of comets (Fig. 25), objects totally different from stars. The simplest hydrocarbon molecule CH is known to exist in the sun and in many other stars of about the same surface temperature. It is present also in comets and in the low-density gas of interstellar space.

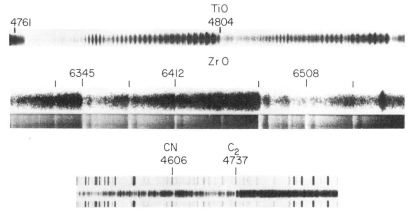

FIG. 23. Spectrograms proving the existence of molecules in stellar atmospheres. Bands of titanium oxide in R Hya; zirconium oxide in T Gem compared with a laboratory spectrum (below) in the red region 6260 A–6600 A; carbon molecules (C_2) 4737 A and CN 4606 A in 19 Psc.

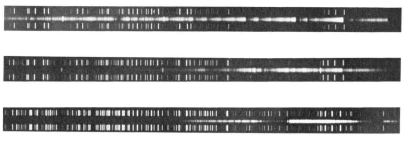

FIG. 24. Molecular bands in variable stars. Titanium oxide in R Cas, type Me; zirconium oxide in R And, type Se; carbon molecules (C_2) in TT Cyg, type Ne.

66   *Space Chemistry*

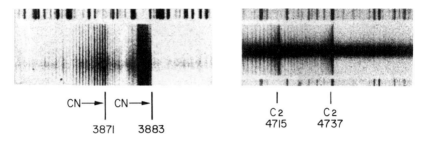

FIG. 25. Emission bands of cyanogen (CN) and carbon (C_2) molecules in the spectrum of Comet Mrkos 1957d.

Molecules containing hydrogen. Because in the universe at large hydrogen atoms greatly outnumber all other kinds we might expect also to find many molecules, H_2, where two hydrogen atoms are bound together in a close partnership. There probably are quantities of these molecules, but unfortunately they do not produce strong spectrum bands in the wavelengths which we can readily observe. Hence our actual knowledge of the location and behavior of hydrogen molecules in the universe is virtually nil. Perhaps future observations in the infrared may remedy the situation.

Molecules containing hydrogen plus another element vary in the strength with which they are bound together. Water vapor molecules, H_2O, are weakly bound and therefore are not found in stars. The molecule CH, more tightly bound, does exist in the stars, as mentioned in a preceding paragraph. Also a few molecules called hydrides, formed of one atom of hydrogen plus one of a metal, are strong enough to hold together in stellar atmospheres. Those observed include hydrides of calcium, magnesium, silicon, and aluminum. These molecules are not abundant and their marks on stellar spectra are inconspicuous. For the most part they can be studied only on high-scale spectra. Aluminum hydride deserves notice for a special reason.

Molecules 67

Nearly all molecules in the stars reveal themselves by absorbing (blacking out) some of the star's light on its way through its atmosphere. This produces characteristic dark spaces in the observed spectrum. Thus aluminum hydride molecules are responsible for certain rather inconspicuous dark markings in spectra of some cooler stars. But in the famous long-period variable star Chi Cygni, something entirely different happens. Unlike the sun and most stars, this huge, cool red star does not stay equally bright at all times. During fairly regular 409-day cycles, it repeatedly exhibits astonishing changes in brightness by a factor of about 1600 (Fig. 26). When the star is faintest, its aluminum hydride molecules glow with their own characteristic rays instead of absorbing some of the light that comes to them (Fig. 27). And most significantly these

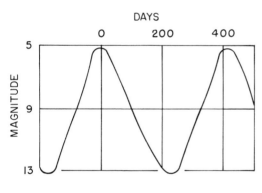

FIG. 26. Light-curves of the long-period variable star χ Cyg.

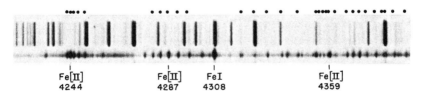

FIG. 27. Identification of the molecular bands of aluminum hydride (AlH) found in emission in χ Cyg at minimum light by G. H. Herbig.

rays are sent out only as the atoms of hydrogen and aluminum are in the act of combining to form molecules. This gives a previously unheard-of and unthought-of opportunity to study a chemical process in progress in a distant star.

These observations will not be easy because Chi Cygni and other stars where similar phenomena occur are very faint and hard to observe at minimum light. In the future, however, such studies will give us valuable information about the temperature and densities of the gases in the outer layers of the cooler stars.

Physicists are now working to obtain better laboratory photographs of the complex spectra of important molecules. They plan to study these in great detail and to publish extensive lists of lines in a form convenient for astronomers to use. This will lead to fuller and more precise knowledge of the presence and behavior of molecules in many astronomical bodies.

V. CHEMICAL SURVEY OF THE UNIVERSE: SOLAR SYSTEM

The underlying fact that the same kinds of atoms and the same physical laws prevail throughout the universe is now well established. This does not mean, however, that all objects have the same chemical composition. The proportions of the various elements differ in various bodies. In some places atoms combine to form molecules, in others they do not. Thus in spite of the fundamental unity of nature, there is a most intriguing variety of chemical differences between various objects in the universe. It is the purpose of this chapter and the following one to describe the more outstanding chemical properties of these objects which may differ so much in size and appearance. We will look first at the various bodies in our solar system; then at the stars, the nebulae, and the materials between the stars; and very briefly at the galaxies.

The sun dominates the solar system in every physical way (Fig. 28). Its mass is 332,000 times that of the earth and 740 times that of all the planets combined. Except for small regions very near the surfaces of each of the planets

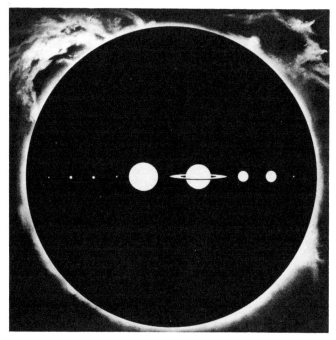

FIG. 28. Comparative sizes of the sun and planets. The large black circle with surrounding prominences represents the sun. The disks of the planets on the same scale are sketched in order within the disk of the sun. From left to right they are: Mercury, Venus, Earth, Mars, Jupiter, Saturn, Uranus, Neptune, and Pluto.

and satellites, its gravitational power is vastly greater than that of the other bodies. The sun has a stronger pull on the moon than does the earth although the sun is 400 times as distant. In a contest of strength the sun would easily take the moon away from the earth. It does not do so because the moon and the earth yield equally to the sun's attraction and travel together in similar, nearly circular orbits about the sun. The earth's small, extra pull deflects the moon's path into a slightly wavy curve, which we self-centered creatures interpret as an orbit around the earth. Actually, the moon's orbit at all points is concave toward the sun.

The energy sent out by the sun in waves of visible and invisible light completely overpowers that sent out, mostly by reflection, by the small, cool planets. At a moderate astronomical distance from the solar system, say at the distance of the nearest fixed star, an observer could see the sun as a point of light just as we see a star, but even with a 200-inch telescope he could detect none of the planets. Despite its physical domination of the solar system, however, the sun is only one of several bodies which attract our interest, and each deserves attention here.

The Earth

The solid portion of the earth is divided by geophysicists into three main parts: core, mantle, and crust, as indicated in Figure 29 and in Table 4.

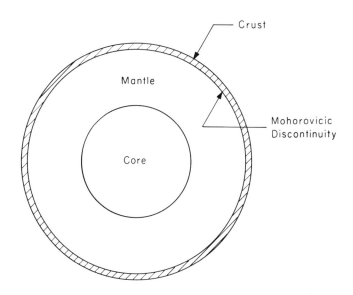

FIG. 29. Sketch of the earth indicating the location of the core and mantle. The crust is very thin. Not drawn to scale.

TABLE 4

Main Structural Features of the Earth

Feature	Radius of Outer Boundary (miles)	Thickness (miles)	Density (water = 1)	Mass (Per cent)
Core	2160	2160	10.7	31.5
Mantle	3940	1780	4.5	67.8
Crust	3960	20	2.8	0.7
Whole earth	3960	3960	5.5	100.

The *core* is composed mostly of iron alloyed with about 8 per cent nickel and possibly with other metals. Its temperature is so high that it would normally be molten, but the pressure and resulting density are so great that it acts like a rigid solid.

The *mantle,* which contains two-thirds of the entire mass of the earth, probably consists mainly of silicates, with the elements oxygen, silicon, magnesium, and iron predominating. A project under study by the Division of Earth Sciences of the U.S. National Academy of Sciences–National Research Council proposes to drill a hole through the earth's crust down into the mantle and to bring up samples for detailed chemical examination. The project is called "Mohole" because it is designed to pierce the "Mohorovicic Discontinuity" which marks the boundary between the relatively thin crust of the earth and the underlying mantle. The Discontinuity was discovered about 1909 by the Yugoslav seismologist A. Mohorovicic from studies of the behavior of earthquake waves. It is about 22 miles below the surface of the continents, but is much thinner under the oceans and thus lies only a very few miles below the ocean floors. Hence it will be strategic to start boring the hole from the bottom of the ocean rather than from a point on land. Preliminary drillings have already been made in 12,000 feet of water near Guadalupe Island, off the western coast of Baja, California. When this project is successfully completed, our knowledge of the chemistry of the earth's mantle will become much more detailed and definite.

The *crust* of the earth is easily accessible to man, and its chemical composition has been carefully studied. In Table 5 the nine elements most abundant in the crust are listed with their relative number of atoms. These nine elements make up 99.25 per cent of the crust. Oxygen and silicon are outstanding. The relative proportions of elements in the mantle, as now estimated, are included for comparison. The crust is much richer in potassium (K), sodium (Na), and aluminum (Al), and somewhat richer in calcium (Ca). It is poorer in iron (Fe) and, especially, in magnesium (Mg).

TABLE 5

Elements in Mantle and Crust of the Earth

	Relative Number of Atoms	
Element	Mantle	Crust
O	39.	30.
Na	0.1	1.2
Mg	16.	0.9
Al	0.4	3.0
Si	10.	10.
K	0.04	0.7
Ca	0.3	0.9
Ti	0.02	0.1
Fe	1.7	0.9

TABLE 6
Composition of Dry Air

Gas	Symbol	Per cent of Volume
Molecular nitrogen	N_2	78.09
Molecular oxygen	O_2	20.95
Argon	A	0.93
Carbon dioxide	CO_2	0.03
Neon	Ne	0.002
Helium	He	0.0005
Methane	CH_4	0.0002
Krypton	Kr	0.0001
Nitrous oxide	N_2O	0.0001
Molecular hydrogen	H_2	0.00005
Ozone	O_3	0.00004
Xenon	Xe	0.000008

The earth's atmosphere. The atmosphere, held down by gravity, extends above the surface of the earth about one-hundredth of the earth's radius. Ninety per cent of it lies below an altitude of 10 miles, 99 per cent below 20 miles, and only one-millionth is above 60 miles. The composition, aside from water vapor, does not vary much with height.

Seventy-eight per cent of the air, by volume, is nitrogen in molecules; 21 per cent is oxygen, also in molecules. Argon is a poor third with 1 per cent; nevertheless the total argon content of the atmosphere is prodigious—*each square mile of the earth's surface supports 800,000,000 pounds of argon*. Near the surface of the earth, water vapor may constitute 1 per cent of the air. At higher altitudes it diminishes rapidly while argon retains nearly the same relative proportion at all altitudes. Hence in the whole atmosphere there is more argon than water vapor. For every million parts of argon there are the following parts of the other inert gases:

Argon	1,000,000
Neon	2,000
Helium	500
Krypton	100
Xenon	1

A more complete listing of the constituents of dry air is given in Table 6. This shows that, other than molecular nitrogen (2 atoms), molecular oxygen (2 atoms), water vapor (2 atoms of hydrogen and 1 of oxygen), and atomic argon, all components taken together comprise only about 3 one-hundredths of 1 per cent of the air. If carbon dioxide (1 atom of carbon and 2 of oxygen) is added to the list, the remaining elements total only three-thousandths of 1 per cent.

Meteorites

Meteorites, or stones that fall from the sky, are of great chemical interest. They are the only bodies from beyond our atmosphere that survive the rapid flight through the atmosphere and are available for detailed laboratory study. Of all the billions of meteors to cross the sky every year, only a tiny fraction reach the earth; the others disintegrate in the air. Approximately 2000 "macrometeorites" large enough to be identified reach the earth each year but of these only about twenty-five are found. The total number of fairly large meteorites (weighing from a few ounces to many tons) in museums today is about 1500.

Meteorites are of two types, iron and stony (Figs. 30 and 31). Iron meteorites are wholly metallic, 90 per cent iron, 9 per cent nickel, and 1 per cent cobalt. Stony

76 Space Chemistry

FIG. 30. Figures in an iron meteorite from South Africa. The crystalline structure is brought out on the cut surface by nitric acid.

FIG. 31. A stony meteorite.

meteorites contain these metals and also a variable proportion of silicates in which the chief elements are:

Oxygen	43%
Silicon	21%
Magnesium	17%
Iron (combined with other atoms)	13%
Calcium	2%
Aluminum	2%

Meteorites do not closely resemble any minerals abundant in the earth's crust but the silicate phase in stony meteorites is similar in composition to the earth's mantle. For this and other reasons, it seems probable that meteorites may be fragments of a planet which was disrupted ages ago. Authorities believe that meteorites are our best source of information on the cosmic abundance of metals, especially of the heavier metals. Many of the figures in the last column of Table 6 depend largely on data from meteorites.

Meteorites are similar to the sun in the relative proportions of the following elements: sodium, magnesium, aluminum, silicon, potassium, calcium, titanium, vanadium, chromium, manganese, iron. This does not prove that meteorites were once ejected from the sun, but it does indicate a common origin for sun and meteorites, and that for the metallic elements both are good samples of general cosmic chemistry.

Tektites

These interesting objects are small rounded glassy bodies made of a substance resembling obsidian. Most of them weigh less than a pound. They are found on the surface of the earth but only in certain regions. The most extensive deposits known are found in the southern part of Australia (Fig. 32) and in the land areas surrounding the South China Sea. They are usually green or yellow and sometimes quite clear. Their surfaces are rounded in such a way as to suggest that they have been heated at least once and probably twice. The second heating may have been caused by passage through the atmosphere. Some are composed of almost pure silica (silicon dioxide SiO_2) and all contain a high proportion of silicates. They contain little water, but otherwise their composition resembles that of sedimentary rocks.

A great variety of hypotheses concerning the origin of tektites has been suggested. They are variously attributed to the action of lightning or of volcanoes, disruption of

FIG. 32. Areas in southern Australia where tektites have been found.

some planet, collisions of the earth with comets; or they may be chips thrown off by the impact of meteors on the earth or the moon. Negative arguments based on the nature of tektites and on their localized distribution on the earth's surface seem to various investigators to show that they could not have been formed on the earth, or on the moon, or farther from the earth than the moon! Recent chemical studies of minor constituents of tektites (such as water, potassium, rubidium, and radioactive isotopes of aluminum and beryllium) appear to geophysicists to preclude an earthly origin. But this seems to be about as far as various investigators agree. Still under discussion is the hypothesis that tektites are fragments of the moon's surface thrown off by impacts of large meteors, and the alternative hypothesis that they are bits of comets that have hit the earth in past times.

Moon

Because the moon is solid and shines only by reflected light, little is known about its chemistry. The spectroscope has told us virtually nothing, for it is difficult to determine the composition of solids by observing them in reflected light. Some reasonable general conclusions, however, can be drawn from observational data. The mean density of the moon is 3.3 times that of water, intermediate between that of the crust and the mantle of the earth. It is thus quite possible that the moon is made of the same materials as the outer portions of the earth. This is one reason why the English mathematician Sir George Darwin suggested that the moon was once part of the earth, that it became detached from the Pacific Ocean area, and has since been slowly receding from the earth. Most authorities now think this improbable.

Some astronomers believe the moon to be largely covered with a thin layer of rather porous material such as dust or volcanic ash. This view is based on the way the moon's surface reflects and polarizes light, and also on observations of temperature which show the surface layers to be poor conductors of heat. It is now generally believed that the numerous "craters" on the moon are not of volcanic origin at all but were caused by the impact of large meteors (Figs. 33 and 34).

FIG. 33. Southern portion of the moon at last quarter, photographed with the 100-inch telescope on Mount Wilson. The numerous craters were probably caused by the impact of large meteors.

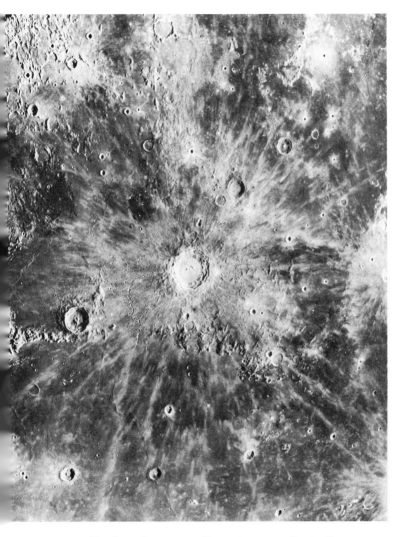

FIG. 34. The large lunar crater Copernicus, 50 miles in diameter, photographed with the 100-inch telescope on Mount Wilson. The bright rays radiating from the crater are best seen at full moon, but their nature and origin are unknown.

In 1958, an interesting observation of the central peak of the lunar crater Alphonsus was made by the Russian astronomer N. A. Kozyrev. He noticed an increase in the brightness of the peak, followed by a sudden change back to normal. A series of spectrograms covering the outburst showed some queer diffuse bright bands near positions of certain well-known bands of molecular carbon. The bands appear somewhat blurred and do not correspond exactly with the carbon bands as observed in the laboratory or in certain red stars. Nevertheless, Kozyrev believes the identification to be correct and that for about thirty minutes the peak emitted a cloud of glowing carbon gas, perhaps as much as 100,000 cubic feet at terrestrial atmospheric pressure. It remains to be seen whether more pseudo-volcanic outbursts of this kind will be discovered on the moon. They are evidently rare.

In a few years when Nova space rockets from the earth begin landing on the moon, either a cleverly designed piece of machinery or a man wearing a pressurized suit and an oxygen helmet will scoop up a cupful of the moon's surface and analyze it on the spot or bring it back to earth for analysis. Then we will know more about the chemistry of the moon.

Mars, Mercury, and Venus

Mercury and Mars have diameters roughly one-half that of the earth, and densities of about $4 \times$ water, compared to 5.5 for the whole earth, 4.5 for the mantle, and 2.8 for the crust. Venus is the earth's twin in size, with a comparable density of 4.9. Because they are similar to the earth in these ways, it is reasonable to believe that these three planets and the earth were formed in much the same way and from the same material.

Mercury keeps so near the sun that the bright background of the sky makes it hard to observe with the spectrograph even when the sun is below the horizon. Hence nothing is known about the composition of its atmosphere.

The atmosphere of Venus contains much carbon dioxide, CO_2, but little oxygen or water vapor (Fig. 35). These facts were discovered in 1932 by W. S. Adams and T. Dunham, Jr., by spectroscopic observations in the infrared made with the 100-inch telescope on Mount Wilson. Until quite recently we have secured little additional chemical information concerning the atmosphere of Venus. The surface of the planet is almost wholly covered by an opaque layer of white clouds; thus the reflected sunlight which comes to us from Venus passes through only the upper part of the atmosphere. The clouds are probably gaseous although it has been suggested that they might be dust clouds whipped up from the dry, relatively hot surface of the planet.

Oxygen and water vapor are hard to detect (Adams and Dunham did not find either) because of the very strong lines of these molecules put into the observed spectrum by the earth's atmosphere. In 1959, however, Dr. John Strong of Johns Hopkins University found definite evidence of water vapor in the atmosphere of Venus. He did it by sending a 16-inch telescope with a recording spectrograph up in a balloon to an altitude of about 70,000 feet where it was above 98 per cent of terrestrial water vapor. The record showed certain infrared water-vapor lines which could be ascribed to Venus.

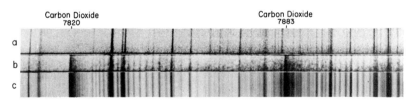

FIG. 35. The spectra of Venus (below) and the sun (above) showing absorption by carbon dioxide (CO_2) in the atmosphere of Venus.

The atmosphere of Mars is thin and usually transparent, yet it contains many mysteries. Yellowish "dust" clouds occasionally appear on the face of the planet, and sometimes white clouds are seen at the edge of the disk. Blue light is scattered just as it is in the earth's atmosphere but even more strongly. Thus photographs taken in blue light ordinarily show no surface markings, only an atmospheric haze (Fig. 36). But for reasons that are not understood the blue haze sometimes disappears for a few days when the planet, as seen from the earth, is in that part of the sky about 180° from the sun. The white polar caps, which observers believe to be thin sheets of ice, come and go with the seasons and it is certain that they are melted or evaporated by the sun's rays.

Careful spectroscopic observations by several observers have failed to disclose any water vapor or oxygen molecules in the Martian atmosphere; if any are present the quantities must be very small. The one constituent definitely identified is carbon dioxide. Spectroscopic observations, in 1952, by G. P. Kuiper of the Yerkes Observatory showed this molecule to be more abundant than in the earth's atmosphere. C. C. Kiess of the U.S. Bureau of Standards has recently announced that his spectroscopic observations indicate the presence of the toxic gas nitrogen peroxide, NO_2, in the atmosphere of Mars. Chemists, however, have expressed doubt.

After decades of conjecture about life on Mars, some chemical evidence supporting a positive view has at last been obtained. Certain areas have long been known to develop a dark greenish color as the Martian spring advances, and it seemed reasonable to attribute the changes to vegetation of some kind. In 1956 and again in 1958 W. M. Sinton of the Lowell Observatory made some significant spectroscopic observations in the infrared (Fig. 37). Working with the 200-inch telescope on Palomar Mountain, he detected, in some of the dark areas of Mars, bands of a CH radical of organic molecules, well known

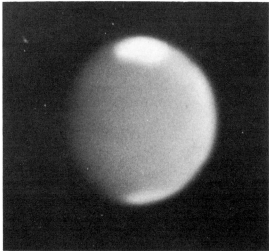

FIG. 36. Mars in red (top) and blue (bottom) photographed with the 60-inch telescope on Mount Wilson in August 1956. Blue light is scattered by the Martian atmosphere.

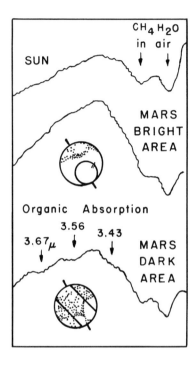

FIG. 37. Evidence for organic molecules in Mars. The curves show intensities in the far infrared spectrum of sunlight reflected from certain areas of Mars. The upper curve (for comparison) is that of direct sunlight which has absorption bands of methane (CH_4) and water vapor (H_2O) present in the earth's atmosphere. In light received from Mars the bright desert area is shown by the middle curve and the dark-greenish area by the lower curve. The dips marked by arrows are absorption bands apparently due to organic molecules on the surface of Mars. The wavelengths are marked in microns, $\mu = 0.001$ mm. The wavelength in angstroms is 36,000.

in light reflected from terrestrial plants. The band was weak or absent in the surrounding bright yellow areas. This means that the organic molecules are on the surface rather than in the atmosphere and are associated with the dark regions.

Sinton stated his conclusion in the following words: "It is tempting to suppose that wherever the planet is dark this band is present, even in regions which have become dark during the history of observation of Mars. If this is true, then organic molecules are produced in localized regions in short spans of time. It would seem that life of some sort, vegetation or possibly even bacteria, would be most capable of doing this."

The Outer Planets

The major planets Jupiter (Fig. 38), Saturn (Fig. 39), Uranus, and Neptune are all much larger than the inner planets discussed above. They are also much less dense. While the density of the earth is 5.5 times that of water, the density of Jupiter, Uranus, and Neptune is 1.3. Saturn is still lighter with a density of only 0.7; it would float if dumped into an ocean big enough to hold it.

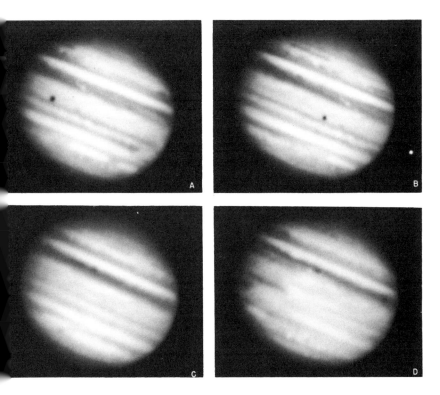

FIG. 38. Four photographs of Jupiter with the 200-inch Hale telescope on Palomar Mountain.

88 Space Chemistry

FIG. 39. Saturn and its ring system photographed with the 200-inch Hale telescope on Palomar Mountain.

The internal constitution of all these planets must be very different from that of the earth. Each may, or may not, have a small metallic nucleus. What corresponds to the mantle of the earth probably consists chiefly of liquefied or frozen gases. The crust may be of frozen water. Careful theoretical work on conditions in the interiors of Jupiter and Saturn indicates that these planets consist largely of hydrogen and helium under sufficient pressure to be liquid or solid. In Jupiter the mass of helium may be about one-fourth that of hydrogen; in Saturn helium may be relatively more abundant. It has been suggested that Uranus and Neptune are composed mainly of water, ammonia, and methane, all in the solid state. These findings of course have not been obtained from direct observation but are based on quite a variety of data and hypotheses. They are by no means certain but probably do represent a fair approximation to the truth.

The atmospheres of the outer planets are thick and dense and not, as on earth, clearly distinguishable from the crust. The markings on Jupiter occur in belts parallel to the equator. They are not rigidly fixed but show relative motion, and although persistent, are not permanent.

Dark bands in the yellow and red have been known for many years in the spectra of sunlight reflected from these major planets. They are strong in Jupiter, stronger still in Uranus and Neptune. It was those of Uranus that Secchi in Italy and Huggins in England detected visually nearly a century ago. Their origin was a mystery until 1932 when Rupert Wildt of Göttingen (now at Yale) showed that they were probably due to methane, a gaseous hydrocarbon, CH_4, and ammonia gas, NH_3 (Fig. 40). The identifications were later confirmed by T. Dunham, Jr., from high-dispersion spectrograms taken with the 100-inch telescope on Mount Wilson. Man could not live in such an atmosphere; he would choke at the first breath. The chances of a man landing on Jupiter and coming back alive are pretty slim. The moon, Mars, and possibly Venus will be more easily conquered.

The methane bands are stronger, the ammonia bands weaker in the outer planets Saturn, Uranus, and Neptune than in Jupiter.

Plausible hypotheses indicate that Jupiter's atmosphere should contain considerable hydrogen, possibly in molecular form. The chief lines of molecular hydrogen lie outside those regions of the spectrum which can be photographed through our atmosphere. Three secondary lines were identified, however, in 1958 by C. C. Kiess of the U.S. Bureau of Standards using a high-dispersion spectrograph on Mauna Loa, Hawaii. Although weak, these lines actually indicate a great abundance of molecular hydrogen in Jupiter's atmosphere.

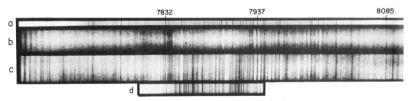

Near Infra-red Spectrum of Saturn and Jupiter
a= Sun; b=Saturn; c=Jupiter; d=Ammonia Gas

FIG. 40. Spectra showing the presence of ammonia (NH_3) faintly in Saturn and strongly in Jupiter.

Comets

These bodies are much larger and more tenuous than any we have yet considered. The diffuse head of a great comet may be more than 100,000 miles in diameter; the tail more than 40,000,000 miles long (Fig. 41). A nucleus at the center of the head may be fairly small, however, perhaps only a few miles in diameter. In 1910, when the earth passed through the midst of the tail of Halley's comet, the head was about 15,000,000 miles away. The density of the tail was so small that its impact on the earth's atmosphere was not felt.

The small mass of a comet is shown in another way. Comets move in elliptical or parabolic orbits under the gravitational attraction of the sun which is at one focus of the orbital curve. The orbit of a comet which passes near a major planet is altered (perturbed is the technical word) by the gravitational pull of the planet, but no comet has ever had sufficient mass to displace the orbit of any planet or satellite by a measurable amount.

The conspicuous head and tail of a comet consist of glowing gases given off by the solid particles in the nucleus under the influence of light, or of high velocity particles, from the sun. Far from the sun the cold cometary material is faint or invisible. F. L. Whipple of the Harvard College Observatory and other students of comets have recently concluded that the nucleus of most comets is not a single very small dense body, but rather a fairly compact group of many small lumps of frozen molecules, mostly methane, CH_4, ammonia, NH_3, and water, H_2O. Mixed in with these is a small proportion of heavy materials, like those in meteorites. This is the "icy conglomerate" hypothesis of the nature of comets.

Near the sun, some of the ices are melted, evaporated, and the molecules partly dissociated. Moreover, the gases become luminous, and therefore detailed and reliable spectroscopic analyses become possible.

AUGUST 22   AUGUST 24   AUGUST 26   AUGUST 27
1957

FIG. 41. Comet Mrkos on four days in August 1957, photographed with the 48-inch Schmidt telescope on Palomar Mountain.

The chief bands in the spectra of the heads of comets are those of the molecules cyanogen, CN, and carbon, C_2 (Fig. 25). Other molecules observed are CH, NH, OH, and C_3. Atomic lines are usually inconspicuous with the exception of the yellow lines of sodium which may become strong when the comet is within 60,000,000 miles of the sun. Lines of iron and nickel were observed in one comet when it was a few million miles from the sun.

The spectacular tails of comets are composed of glowing molecules streaming rapidly away from the head. The physical behavior and the chemistry are complex. Many types of molecules including CN, C_2, CO, CO_2, N_2, CH, and OH have been identified. In some comets a straight tail points directly away from the sun while a curved plume lags behind as the head of the comet moves along in its orbit. The straight tail is more tenuous than the curved one, and its molecules travel faster away from the

92 *Space Chemistry*

head. Moreover, its composition seems to differ, the molecules CO and N_2 being especially prominent.*

The chemistry of comets and the physical causes of their exotic behavior are still only partially understood. Astronomers are now giving renewed attention to the composition, structure, and probable origin of these strange objects, so different from all other astronomical bodies. One reason for this renewed interest is the hope that a fuller knowledge of comets may eventually shed light on the chemical composition of interstellar dust.

Meteors

Vast numbers of faint meteors can be seen in our upper atmosphere on dark nights (Fig. 42). Many of these occur in showers on about the same date every year, such as the Perseids which enliven the sky every year on August 11. They are called Perseids because their apparent paths through the sky appear to diverge from a point in the constellation Perseus. The divergence is not real but due to perspective, for meteors actually move in parallel lines, continuing the common orbital motion with which they strike the earth's atmosphere. Their orbits can be calculated by straightforward astronomical methods. Many coincide with the orbits of old, worn-out comets. The Perseid meteors, for example, are still pursuing the same orbit followed by Comet 1862 III.

No meteorites have ever been recovered from the Perseids or any other meteor shower. This is understandable if the meteors in showers are formed of icy conglomerates as are comets, for they would be largely vaporized by their

---

*A detailed review of our knowledge of the complicated chemistry of comets, with many photographs of their spectra, may be found in an *Atlas of Representative Cometary Spectra* by P. Swings and L. Haser, published in 1956 by the University of Liège, Belgium.

FIG. 42. The path of a meteor which passed across the field during the exposure on the irregular nebula NGC 281 on August 11, 1921.

high-speed flights through the air. Any remaining solids would presumably be in the form of tiny grains.

Nevertheless, some information concerning the chemical composition of meteors can be obtained from photographs of their spectra. Meteors flash through the sky so quickly and so unexpectedly that it is impossible to turn a telescope on them. But if, on the night of a meteor shower, a wide-angle camera with a prism in front of its lens is kept pointing upward, a spectrum photograph will occasionally be obtained. A few hundred small scale spectra have now been accumulated.

Nearly all meteor spectra have lines of iron and of calcium. In addition, sodium, magnesium, aluminum, silicon, chromium, manganese, and nickel have been reported in certain meteors. The relative prominence of various lines probably depends on the speed with which the meteors move through the air and possibly on the altitude at which they are observed. The presence of molecular bands is not certain and requires further study. It seems probable, however, that comets and meteors are similar if not identical in composition. The spectra of the two bodies look quite different, mostly because they become luminous in quite different ways. Meteors are heated intensely and caused to shine by their rapid motion through the air, while comets are made to glow by fluorescence by light rays or by fast-moving particles shot out by the sun.

Sun

The sun gives us a chance to study a star more closely. Although from a cosmic point of view it is only one of millions of small stars in a rather ordinary galaxy, it is the only one that appears to us larger than an optical point. On no other star can we see any details such as dark spots, bright prominences and flares, or any closely surrounding halo (corona). We are fortunate in having the sun as a sample star, favorably placed for detailed observation (Fig. 43).

Even though the sun appears as a large disk, and we can study all sorts of interesting details on its surface, there remains an important limitation (which applies also to the stars). We cannot probe the depths of the sun but must limit our observations to the outermost layers, which contain only a tiny fraction of 1 per cent of the sun's whole

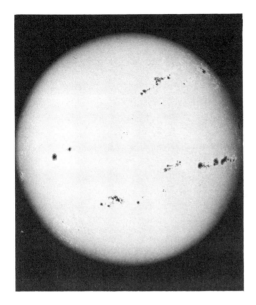

FIG. 43. Direct photograph of the sun at time of sunspot maximum, December 21, 1957.

mass. This is because nearly all of the sun is opaque. The visible surface of the sun is called the "photosphere" and is very near the outside. Above it lies a thin layer of gas called the "reversing layer" where the dark lines of the sun's spectrum are formed. It derived its name from the fact that the atoms in it absorb their characteristic spectrum lines. These lines then appear dark against the glowing background spectrum of light from the photosphere. The photosphere gives out a brilliant and steady stream of light from which the reversing layer subtracts a little in the form of these all-important spectrum lines (Fig. 44). The corona subtracts or adds so little that, except during an eclipse, it can be detected against the bright background of our sky only by very specialized methods. It is really the reversing layer that we are talking about

96 *Space Chemistry*

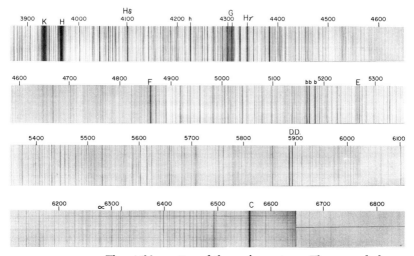

FIG. 44. The visible portion of the sun's spectrum. The many dark lines may be identified by comparing their positions with those of elements observed in the laboratory, and thus the composition of the solar atmosphere is determined. Wavelengths in angstroms are marked above each strip. Photographed with the 13-foot spectrograph at the Mount Wilson 60-foot tower.

when we speak of the chemical composition of the sun. How good a chemical sample of the whole sun this thin reversing layer is depends on how well the materials of the interior of the sun are mixed with those on the surface. Motions within the sun cannot be observed and are difficult to calculate. Astronomers believe that this layer represents quite well the composition of at least the outer half of the sun's mass.

Sunspots are transient dark markings that appear in certain zones on the sun (Fig. 45). They are never seen at the poles, and seldom near the equator. They appear in greatest numbers at intervals of ten or eleven years (Fig. 46) and exhibit fairly strong magnetic fields. Their general behavior is well known but the physical explanation is still incomplete.

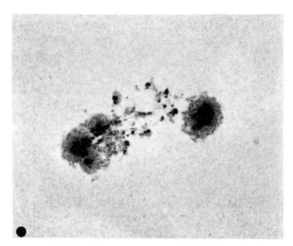

FIG. 45. Enlarged sunspot group. The black circle shows the size of the earth on the same scale.

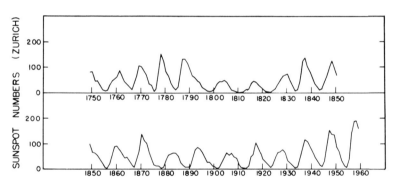

FIG. 46. Curve of Zurich sunspot numbers for 210 years from 1750 to 1960, showing the 11-year period of spot activity.

Table 7, compiled by Dr. Charlotte M. Sitterly, a leading authority on the solar spectrum, lists the elements that are known to exist in the sun. Their proportions, to the best of our knowledge, are about the same as those of the universe as a whole. (See Table 1.)

TABLE 7
Elements in the Sun

Atomic No.	Symbol	Element	Atomic No.	Symbol	Element
1	H	Hydrogen R	40	Zr	Zirconium R
2	He	Helium*	41	Nb	Niobium R
3	Li	Lithium*	42	Mo	Molybdenum R
4	Be	Beryllium R	44	Ru	Ruthenium
5	B	Boron*	45	Rh	Rhodium R
6	C	Carbon R	46	Pd	Palladium R
7	N	Nitrogen	47	Ag	Silver R
8	O	Oxygen	48?	Cd	Cadmium R
9	F	Fluorine*	49	In	Indium*
10	Ne	Neon*	50	Sn	Tin R
11	Na	Sodium R	51	Sb	Antimony
12	Mg	Magnesium R	56	Ba	Barium R
13	Al	Aluminum R	57	La	Lanthanum R
14	Si	Silicon R	58	Ce	Cerium R
15	P	Phosphorus	59	Pr	Praseodymium
16	S	Sulfur	60	Nd	Neodymium R
18	A	Argon*	62	Sm	Samarium
19	K	Potassium R	63	Eu	Europium
20	Ca	Calcium R	64	Gd	Gadolinium
21	Sc	Scandium R	65	Tb	Terbium
22	Ti	Titanium R	66	Dy	Dysprosium
23	V	Vanadium R	68	Er	Erbium R
24	Cr	Chromium R	69	Tm	Thulium
25	Mn	Manganese R	70	Yb	Ytterbium
26	Fe	Iron R	71	Lu	Lutecium
27	Co	Cobalt R	72	Hf	Hafnium
28	Ni	Nickel R	73?	Ta	Tantalum
29	Cu	Copper R	74	W	Tungsten
30	Zn	Zinc R	76	Os	Osmium
31	Ga	Gallium	77	Ir	Iridium
32	Ge	Germanium R	78	Pt	Platinum
37	Rb	Rubidium*	79?	Au	Gold
38	Sr	Strontium R	82	Pb	Lead R
39	Y	Yttrium R	90?	Th	Thorium

TABLE 8

Elements NOT *Detected in the Sun*

Atomic No.	Symbol	Element	Atomic No.	Symbol	Element
17	Cl	Chlorine	54	Xe	Xenon
33	As	Arsenic	55	Cs	Cesium
34	Se	Selenium	61	Pm	Promethium
35	Br	Bromine	67	Ho	Holmium
36	Kr	Krypton	75	Re	Rhenium
43?	Tc	Technetium	80	Hg	Mercury
52	Te	Tellurium	81	Tl	Thallium
53	I	Iodine	83–102*		

*With the possible exception of 90 Th Thorium.

Table 8 lists those elements whose lines have *not* been recognized in sun spectra. This does not necessarily indicate that the elements are completely absent. In fact, nearly all are probably present although in relatively small amounts. Their spectral lines may be numerous but very weak and hard to identify, or the chief lines may be in a part of the spectrum not yet thoroughly studied.

The spectrum of a sunspot differs considerably from that of the normal disk on either side (Fig. 47). There are

---

*Notes to Table 7
 2. Helium. In chromosphere (upper part of reversing layer) and prominences.
 3. Lithium. Evidence supplied by sunspots.
 5. Boron. Detected only in the molecule boron hydride.
 9. Fluorine. Detected only in the molecules magnesium fluoride and strontium fluoride.
 10. Neon. Detected very recently by observations from rockets.
 18. Argon. Evidence supplied by the corona.
 37. Rubidium. Evidence supplied by sunspots.
 49. Indium. Evidence supplied by sunspots.
 R The elements marked R were included in a list published in 1891 by Henry A. Rowland of Johns Hopkins University.

many additional lines and changes in intensity, because the gases in the spot are cooler than those of the normal reversing layer. More molecules are formed at the lower temperature and they account for most of the new lines appearing in the sunspot spectrum. Even in lines due to individual atoms, marked effects of temperature are present. At the lower temperature, lines from normal or undis-

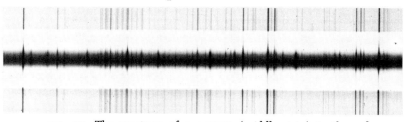

FIG. 47. The spectrum of a sunspot (middle strip) in the red, 6260 A–6320 A. Many more lines are present in the spot than in the hotter solar spectrum above and below.

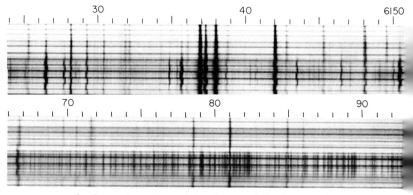

FIG. 48. Sunspot spectrum. The zigzags of the lines as they cross the spot result from reversing the orientation of the alternate analyzing elements which are placed just in front of the slit of the spectrograph. The strength of the magnetic field at any point can be found by measuring the displacements (above) 6124 A–6150 A. (Below) 6366 A–6392 A, numerous fainter lines of the spot which do not show in the solar spectrum of the disk on either side are components of the molecular bands of calcium hydride (CaH). Photographed with the 75-foot spectrograph at Mount Wilson.

turbed atomic states are strengthened, those from high-energy states into which atoms are jostled by collisions with other particles are weakened.

In the spot gases, some of the lines are widened by the presence of magnetic fields. This widening is called the Zeeman effect after the Dutch physicist who discovered it in the laboratory. The zig-zag, crankshaft appearance of the lines in Figure 48 is artificial; it is caused by polarizing strips placed over the slit of the spectrograph. The magnetic field not only widens the lines but polarizes their two edges differently.

Thus the physical conditions in spots differ decidedly in two ways—lower temperature and the presence of a magnetic field—from those of the surrounding solar gases. And more molecules are present; but the actual proportions of the individual atoms of various kinds, counting together the separate atoms and those combined in molecules, are doubtless nearly identical. These conclusions are interesting in themselves, but they also help us in studying sequences of stars at different surface temperatures.

Prominences, spectacular clouds of glowing gas, move like fireworks above the surface of the sun. Often they are projected upward like fiery geysers (Figs. 49 and 50). Their quick changes are probably due to local magnetic fields near the sun's surface, but their queer and rapidly varying motions are not clearly understood.

Prominences are probably composed of ordinary solar gases. Because they are much hotter than the reversing layer, their spectrum exhibits considerable differences from that of the disk. At the sun's limb the prominences appear in emission but on the disk they are dark (Figs. 51 and 52). It was in the spectrum of a prominence that helium was first discovered. The strongest spectral lines are those of calcium, hydrogen, and helium. Their intensities correspond roughly to the characteristic chemistry of the universe—better than do those of the dark lines in the usual solar spectrum.

102   *Space Chemistry*

FIG. 49. Prominences at the limb of the sun.

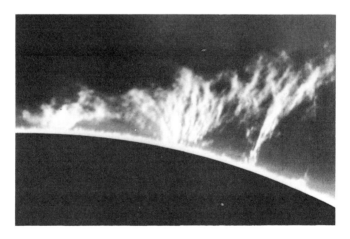

FIG. 50. Large solar prominence of August 21, 1909.

Solar System 103

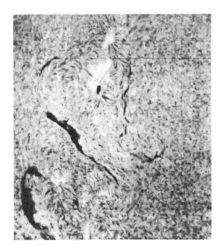

FIG. 51. Prominences projected on the disk of the sun as irregular dark streaks in hydrogen (Hα) light.

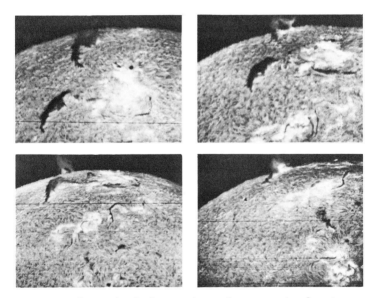

FIG. 52. Hydrogen (Hα) photographs on four successive days in 1917, June 27–28 (above), June 29–30 (below), showing solar features carried over the limb by solar rotation. The prominences appear in emission outside the disk of the sun.

104   *Space Chemistry*

The *corona* is a huge pale green halo surrounding the sun, visible only at times of total eclipse, that is, when the moon fits neatly over the sun's disk, shutting out the dazzling sunlight and darkening the sky. Most people are familiar with the corona through many fine eclipse photographs (Fig. 53). Recently astronomers have developed specialized methods of observing the corona without an eclipse, from high altitude stations where the sky is relatively dark.

The gas of the corona is extremely tenuous but the temperature as measured by the motions of the particles is extremely high—at least a million degrees centigrade. The source of this temperature is believed to lie in shock waves transmitted upward through the photosphere and reversing layer. The shock waves die out in the gas of the corona but their energy goes into high heat motions of the particles.

A peculiar type of spectrum is emitted as a result of the high temperature. For many years it was completely enig-

FIG. 53. The corona of the sun photographed at Green River, Wyoming, June 8, 1918, at the time of total eclipse. The temperature of the corona is extremely high but it probably has the same chemical elements as the sun itself.

matic, and a hypothetical element, coronium, was invented to explain it. We now account for the strange lines by ordinary atoms, but at a temperature too high and a density too low to study in the laboratory.

Hydrogen and helium are doubtless present in the corona but their electrons are all stripped away by the high temperature, and the bare nuclei have no way of producing spectrum lines. In the heavier elements, a dozen or so electrons are stripped away but the remaining ones yield lines that were identified recently by physicists, notably W. Grotrian and B. Edlen. Table 9 lists the atoms that have been identified and the number of electrons missing from the normal complement.

TABLE 9
Atoms Identified in the Solar Corona

Element	Atomic No.	No. of Electrons Missing
Argon	18	9, 13?
Calcium	20	11, 12, 14
Iron	26	9, 10, 12, 13, 14
Nickel	28	11, 12, 14, 15

A novel method of taking photographs of the sun in the light of a single element was invented about 1890 independently by G. E. Hale in the United States and by H. Deslandres in France. A large image of the sun is projected on the slit of a special spectrograph called a spectroheliograph. The resulting spectrum of a strip of the sun falls on a diaphragm with a narrow opening through which passes a single line formed by a particular element. Thus an image of one narrow strip of the sun is obtained. Then the slit is made to move across the sun's image while the photographic plate behind the diaphragm is moved at the

106   *Space Chemistry*

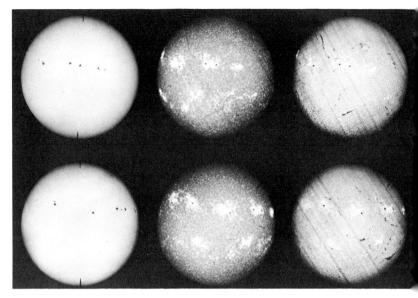

FIG. 54. Chemical photographs of the sun. Upper row; October 10, 1926: left, direct photograph in white light; middle, in violet light of K (calcium); right, in red light of hydrogen (Hα). Lower row; the same four days later. The changes in position of the markings are due to the sun's rotation.

same rate, until a picture of the whole sun is obtained. In this fairly simple way, a marvelous photograph is obtained which shows the behavior of a selected chemical element on the sun's surface. It is as if one could take a picture of a scrap heap in such a way that only one metal, iron or tin or copper, would show. The appearance of the sun is quite different from that of an ordinary solar photograph taken directly in the focus of the telescope lens without the formation of a spectrum. Examples are shown in Figure 54. The brightness of the vapor of an element at any point on the sun's surface depends on the temperature and density at that point. Hence these photographs supply much physical data for a detailed study of the complicated phenomena on the sun's surface.

VI. CHEMICAL SURVEY OF THE UNIVERSE:
STARS AND NEBULAE

Stars

The chemistry of the stars is an enormous subject. Several million stellar spectra have been recorded, mostly on photographs of star fields taken through a prism in front of the lens of a medium-sized telescope. For several hundred thousand stars the type of spectrum has been systematically listed, and some 16,000 stars have been more fully analyzed one at a time by focusing them on the slit of a spectrograph placed at the eye-end of a large telescope. The spectrum photographs then taken show numerous details, each of which has a chemical interpretation. Comparison spectra, usually of iron lines from an electric arc, are photographed close on either side of the star spectrum and make it possible to determine accurate wavelengths and chemical identifications of the features seen in the star's spectrum. This is a well-established routine at various observatories and long research papers record the chemical composition of the atmospheres of individual stars.

The first systematic classification of stars was based on their apparent chemistry as revealed by the dominant lines in their spectra. Father Secchi's four main types of stellar spectra can now be described chemically as follows: (I) strong hydrogen lines; (II) numerous metallic lines; (III) bands of titanium oxide; (IV) bands of carbon and cyanogen. These types are in order of decreasing stellar temperature, with Types III and IV covering about the same range.

The need for a more detailed classification was met about 1890 by the Harvard system, which is now in use generally. The early history was briefly outlined in Chapter II, but since the Harvard Classification is now so widely used in discussions of the chemistry of the stars as well as of their spectra, we ought to take a somewhat closer look at it.

In Figure 55, the Harvard classes are laid out from left to right in order of decreasing temperature. (The temperature used is the apparent surface temperature of the stars.) The corresponding range of color is from bluish, through white and yellow, to red. The vast majority of all stars fall in these classes. The curious alphabetical sequence O, B, A, F, G, K, M came about as follows. In the first attempt at classification, stars with the strongest hydrogen lines were called class A, while others were arranged in order of decreasing prominence of these lines and were called B, C, D, E, ... M in normal alphabetical order.

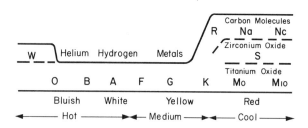

FIG. 55. The sequence of Harvard spectral types according to temperature and the lines observed in stellar spectra.

The system was soon simplified by cutting down the number of groups, for some of them turned out to be unnecessary or due to faulty photographs. In this way classes C, D, E, H, I, J, and L were eliminated. Then class B was examined more closely. Although their hydrogen lines are weaker, stars in this group were bluer than the A's and had few lines of metals. It soon became clear that these bluer stars are hotter than the A's; hence in the adopted sequence of temperatures, B was placed first. Later, still hotter stars were found and had to be given the letter O, far down the alphabet. This is the story of the queer alphabetical sequence O, B, A, F, G, K, M (Oh, be a fine girl. Kiss me.) now universally used by astronomers as a kind of shorthand in stellar chemistry. The general appearance of stellar spectra of various classes is shown in Figure 56.

The surface temperature determines the color of a star in the same way that the temperature of a metal rod in a furnace determines its color. In the furnace, a rod heated just enough to give out any perceptible light has a deep red color. As it becomes hotter, its color changes to a lighter red, to yellow, and then to white. Stars exhibit exactly the same sequence but continue it somewhat further, with the very hottest stars looking bluish. A pair of bright visible stars in the constellation Orion exhibit colors near the extremes of this color-temperature sequence. They are the relatively cool red Betelgeuse, Alpha Orionis, and the hot bluish Rigel, Beta Orionis.

The hydrogen stars (class A), while much hotter than the sun (class G), are exceeded in temperature by both B and O stars. Temperatures of the various classes are listed in Table 10. They are given on a special scale regularly used in physics and astronomy called the absolute scale, which is like the usual centigrade scale except that its zero point is 273° lower. This change of zero simplifies physical calculations by putting temperature on an energy basis. Atomic motions cease and the energy of their motion becomes zero at 0° absolute.

110 *Space Chemistry*

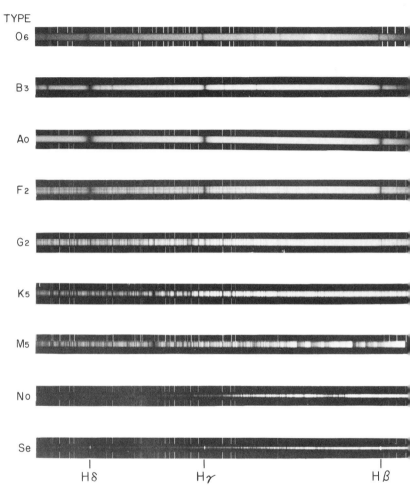

FIG. 56. Principal types of stellar spectra. In the white, hot stars of types O, B, and A, hydrogen lines are strong. The cool red stars of types M, S, and N are distinguished by bands produced by molecules. The intermediate types, F, G, and K, have numerous lines of metals.

TABLE 10
Surface Temperatures of Stars

Spectral Class	Dominant Spectral Features	Surface Temperature in Centigrade Degrees Absolute
O	Ionized helium	80,000°
B	Neutral helium	20,000
A	Hydrogen	10,000
F	Intermediate between A and G	7000
G	Metals	6000
K	Intermediate between G and M	4200
M	Titanium oxide	3000
S	Zirconium oxide	3000
N	Carbon molecules	3000

The temperatures listed in Table 10 refer to the photospheres of the stars and correspond to the colors seen by a distant observer. Just as in the sun, a thin layer of gas, the so-called reversing layer immediately above the photosphere, produces the characteristic spectrum lines; it is in contact with the photosphere and has nearly the same temperature.

It has long been known that lines of different elements predominate in stars of different colors. It was not easy to believe, however, that stars at various temperatures have totally different chemical compositions. The answer to this fundamental problem came slowly, but is now almost complete.

In laboratory experiments with the electric furnace Arthur S. King of the Mount Wilson Observatory showed clearly that the relative intensities of various spectrum lines of a metallic element such as iron or titanium depend on the temperature of the gases emitting or absorbing them. Previously, physicists had thought that differences in electrical excitation were necessary to produce differences among the lines of any element. In a very few years the theory of chemical equilibrium together with the newly developed quantum theory of atomic structure gave a consistent, although revolutionary, explanation of the general behavior of the lines of various elements at different temperatures. Application of this theory to stellar spectra makes it probable that typical stars of classes O, B, A, F, G, K, M all have the same chemical composition.

It can be generalized that nearly all stars—whatever their color or whatever lines may appear in their spectra—have virtually the same chemical composition. But this is not the whole story. We now turn to important examples of departure from this rule of uniform composition.

First notice in Figure 55 that the right-hand, low-temperature end of the sequence divides into three branches marked M, S, and N. This means that there are three kinds of red stars whose spectra differ in definite and characteristic ways. Of these three types those designated M are by far the most numerous. They have the same chemical composition as the hotter stars but are cool enough to allow some metal atoms, notably those of titanium, to unite with oxygen atoms to form oxide molecules. The spectra of stars of classes S and N differ definitely from those of class M, but because the three classes have about the same temperature, we cannot, as in the chief sequence of classes O to M, explain the differences in terms of differing temperatures. Instead we must conclude that there the differences in spectral lines represent real chemical differences, and that there actually are three different kinds of red stars. This is a conclusion of considerable importance. (See Figures 23 and 24.)

The spectra of class S stars have some resemblance to those of class M but the titanium oxide bands are weak or absent; instead, the characteristic features are bands due to zirconium oxide. Also there are slight but significant differences in the rich array of dark lines due to atoms. All the evidence indicates a greater abundance of the heavier metallic elements, notably strontium, atomic number 38, yttrium 39, zirconium 40, niobium 41, technetium 43 (see Chapter III), and barium 56. Also especially abundant are the heavy "rare earth" elements such as neodymium 60, samarium 62, europium 63, and gadolinium 64. Nuclear physicists are discussing what circumstances could lead to a greater abundance of heavy elements in atmospheres of S-type stars, but the problems are difficult. The characteristic presence of technetium, an unstable element with a natural half-life of only 200,000 years, furnishes a significant clue which has not yet been interpreted with certainty. We believe that all these heavy elements are formed in the hot central regions of stars but we do not know how long it takes for elements formed in the opaque interior to diffuse outward into the atmosphere where they can produce observable spectrum lines. The possibility that technetium and perhaps other heavy elements may be formed right in a star's atmosphere by impact of neutrons or even of cosmic rays has been considered but no firm conclusions have been reached.

Red stars of the third kind, class N, often called carbon stars, are entirely different. Their spectra are characterized by strong bands of molecules containing carbon, the two most conspicuous groups of bands being those due to diatomic carbon, C_2, and cyanogen, CN. Bands of metallic oxides are absent. Compared to other kinds of red stars the atmospheres of these N stars must have a preponderance of carbon atoms.

The molecules that dominate the spectrum of a red star seem to depend on the balance of the numbers of carbon and oxygen atoms in the star's atmosphere. An oxygen atom having its choice of combining with an atom of carbon or with a metal atom will choose the carbon atom as its partner. If there were an equal number of carbon and oxygen atoms, all would come together to make carbon monoxide molecules, CO, and no other kinds of molecules containing either carbon or oxygen would be formed. A few cool red stars like R Canis Minoris in whose spectra bands both of metallic oxides and of carbon compounds are virtually absent seem to represent this equality of carbon and oxygen. Carbon monoxide should be abundant, but unfortunately it has no outstanding spectral features in the observable portions of the spectrum; hence we have no direct observational check on the hypothesis of equal abundance.

If carbon atoms outnumber those of oxygen, the excess atoms will as second choice combine either with other carbons forming C_2 or with atoms of nitrogen forming CN, cyanogen. These two kinds of molecules are responsible for the characteristic carbon bands in spectra of class N.

If, on the other hand, oxygen atoms outnumber those of carbon, the excess oxygens will seek metallic atoms as partners. Apparently their first choice among the metals is zirconium with which they will combine to form zirconium oxide, which produces the bands characteristic of class S. But if, as in most red stars, zirconium atoms are scarce, the oxygen atoms will wait until the temperature is somewhat lower and will then accept the much more abundant titanium atoms as partners, forming molecules of titanium oxide, which produce the well-known bands in spectra of class M.

The considerations in the two preceding paragraphs are indicated in the diagram shown in Figure 57. The main co-ordinate in the description of star spectra is, of course, temperature. Plotted in a plane at right angles to the temperature axis are the two chemical ratios we have just been discussing. In this plane the horizontal co-ordinate is the ratio of carbon to oxygen. Stars with equal numbers of carbon and oxygen atoms would lie on a vertical line through the middle of the upright plane. The vertical co-ordinate is the ratio of heavy metals to those of the iron group; for simplicity we may say the ratio of zirconium to

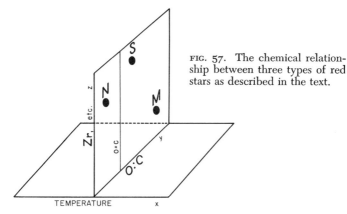

FIG. 57. The chemical relationship between three types of red stars as described in the text.

titanium. The appropriate places for the three spectral classes are marked in the diagram. Class M with much oxygen and little zirconium is low to the right. Class S with nearly as much oxygen but with much more zirconium is nearly as far to the right but toward the top. Class N with much less oxygen than carbon lies far to the left and rather high because the number of zirconium atoms is considerable.

We have seen that, according to their chemistry, red stars naturally fall into three groups. Classified according to size and brightness, however, they divide into two main groups, giants and dwarfs. Most conspicuous in the sky are the red giants, huge objects of low density, hundreds or even thousands of times as bright as the sun. Most of them are of class M, with the standard chemical composition, but a small fraction are in the divergent S and N classes. The faintest red dwarfs, condensed objects, some of them less than one-thousandth as bright as the sun, are, as far as we know, all of class M. Thus all dwarfs seem to build up the chemical elements according to a standard pattern. It is now believed that the dwarfs may at a certain stage of their evolution develop into giants, but exactly

116 *Space Chemistry*

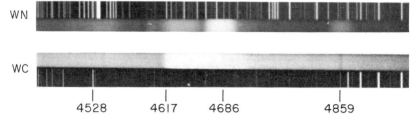

FIG. 58. The peculiar spectra of two hot Wolf-Rayet stars. One (below) HD193793, WC, has strong wide emission lines of carbon; the other (above) HD192163, WN, has more nitrogen than carbon. The explanation is yet to be found.

what circumstances cause a few of the giants to deviate from the standard chemistry we do not yet know.

At the high-temperature end of the stellar sequence is another quite different example of chemical divergence. The very peculiar Wolf-Rayet stars, marked W in Figure 55, are still not well understood. Their characteristic spectrum lines are bright instead of dark, and are very wide, probably indicating rapid atmospheric motions. One thing we know is that the gases emitting the lines are at a very high temperature. Also it seems clear that there are two chemical groups within the same temperature range, one, WC, having plenty of carbon with little nitrogen, the other, WN, having more nitrogen than carbon (Fig. 58). The explanation is yet to come.

A small group of class A, or hydrogen-line, stars with a most interesting chemical peculiarity deserves special mention. In this group of so-called "peculiar A" stars, lines of a number of the "rare earth" elements from cerium, atomic number 58, to lutetium, 71, are unusually intense, lines of europium being especially prominent. H. W. Babcock of the Mount Wilson and Palomar Observatories recently discovered that the edges of certain lines in the spectra of many (perhaps all) of these stars are polarized. This means that the gas producing the lines lies in a fairly strong magnetic field. In most of the stars observed both the lines and the magnetic fields are variable in intensity. Neither the observations nor the theory are easy, but these chemically peculiar stars may eventually yield much valuable information about stellar atmospheres.

Among the hundreds of thousands of stars whose spectra have been photographed are a few whose chemistry seems weirdly different from the others. For example, a few stars seem deficient in hydrogen or to have more than their share of helium. One odd "white dwarf" studied by J. L. Greenstein would, if we take naively the indication of its spectrum lines, consist almost wholly of magnesium and calcium (Fig. 59). This is not actually true, but the circumstances that cause this exceptional spectrum are not well understood.

Stars have recently been divided into two groups (called "Populations" I and II) according to the abundance of their metallic atoms. In stars in certain globular clusters, the relative number of metallic atoms is decidedly less

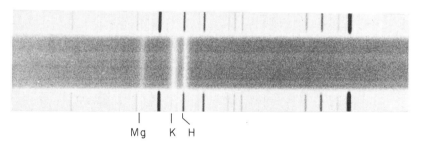

FIG. 59. The very unusual spectrum of the "white dwarf" star Ross 640. The two strong lines are H and K of ionized calcium. The fainter line to the left is due to neutral magnesium, a blend of 3832 A and 3838 A. Other chemical elements are doubtless present in the star's atmosphere but conditions are such as to suppress their spectral lines. The comparison lines above and below are helium and hydrogen.

than in stars such as the sun; in an extreme case, in certain bright stars in the cluster M 92 (see Fig. 75), the relative abundance of metals may be only one one-hundredth that in the sun. Other groups of stars exhibit, somewhat less strikingly, the same kind of deficiency. There is further discussion of this subject and its probable significance in stellar evolution in Chapter VII.

Nebulae

Soon after telescopes were put into use observers discovered small faintly luminous patches, different from either stars or planets. These were called "nebulae"—after the Latin word for cloud—and were of two kinds. Some were diffuse with ill-defined boundaries. Others, especially the smaller ones, had well-defined circular or elliptical outlines and looked somewhat like faint planets; for this reason they were called planetary nebulae. The nature of both kinds was unknown but the best guess was that they were distant unresolved groups of stars. This was in analogy with the fact that to the naked eye the light from the individual stars in the Milky Way seems to coalesce to form continuous luminous clouds.

This was the status of nebulae when Sir William Huggins began observing more than a century ago. His first observations were "direct," that is, ordinary telescopic views showing brightness, form, and color, but he soon turned to the spectroscope to learn more about the nature of heavenly bodies. Here is his own account of his first spectroscopic examination of nebulae:

"On the evening of August 29th, 1864, I directed the telescope for the first time to a planetary nebula in Draco.

"I looked into the spectroscope. No spectrum such as I expected! A single bright line only! At first I suspected some displacement of the prism, and that I was looking at a reflection of the illuminated slit from one of its faces.

This thought was scarcely more than momentary; then the true interpretation flashed upon me. The light of the nebula was monochromatic; and so, unlike any other light I had as yet subjected to prismatic examination, could not be extended out to form a complete spectrum. After passing through the two prisms it remained concentrated into a single bright line, having a width corresponding to the width of the slit, and occupying in the instrument a position at that part of the spectrum to which its light belongs in refrangibility. A little closer looking showed two other bright lines on the side towards the blue, all the three lines being separated by intervals relatively dark (Fig. 60).

"The riddle of the nebulae was solved. The answer, which had come to us in the light itself, read: Not an aggregation of stars, but a luminous gas. Stars after the order of our own sun, and of the brighter stars, would give a different spectrum; the light of this nebula had clearly been emitted by a luminous gas."

FIG. 60. Two views of the planetary nebula NGC 6543 in Draco, and its spectrum. In 1864 Huggins learned from its spectrum that the nebula had only emission lines and hence was entirely gaseous. Two lines of [OIII] are at the extreme right.

Huggins' solution of the riddle of the nebulae was only partial, as he himself hastened to point out, for the outstanding spectral line which he had discovered was not matched (and has not yet been matched) by any produced in the laboratory. The chemical nature of the gas that emits this strange line had yet to be discovered.

Knowledge of the spectral lines of nebulae advanced rapidly after Huggins successfully applied photography in 1882. Relatively faint nebular lines were eventually identified with hydrogen, helium, carbon, nitrogen, and oxygen; but still no counterpart could be found for the original green line or for its companion line discovered later. Thus it came to be believed that these lines were produced by a substance not found on earth which, in anticipation of its acceptance as a bona fide chemical element, was named "nebulium."

This belief was seriously questioned when the integral structure of the table of chemical elements became known, and it appeared that all possible light-weight gases were already accounted for. This development forced the idea that the strange lines came not from a special celestial element, but from a common element shining under physical conditions not duplicated in the laboratory. When physicists had accumulated sufficient knowledge of atomic structure to compute with reasonable accuracy the wavelengths of spectral lines which atoms might emit under various circumstances, the stage was set for the second discovery concerning nebulae. Dr. Ira S. Bowen began to wonder what spectral lines common atoms might emit if left to themselves for relatively long intervals of time. In the laboratory, even in a so-called vacuum, an atom can go scarcely 1/1000 of a second without colliding with another atom or with the walls of the container; but in the much lower densities of nebulae in the depths of space, the intervals between atomic collisions could be hundreds or even thousands of seconds. Making some calculations from well-established data on various atoms, Bowen soon was

Stars and Nebulae 121

able to prove that the green lines characteristic of nebulae are produced by oxygen—plain old earthly oxygen such as men have been breathing for a long time.

The oxygen atoms that enter human lungs differ from those in nebulae in certain important particulars, depending on the situation in which they are placed, but not at all in their fundamental internal structure. We breathe *molecular* oxygen—two atoms closely bound together. In nebulae, oxygen exists only in the *atomic* state. Moreover, some of the outer electrons of the single atoms are knocked off by the impact of extreme ultraviolet light from the hot central star. The green lines discovered by Huggins are in fact produced by oxygen atoms each of which has lost two of its complete complement of eight electrons.

Gaseous nebulae are of two kinds: (1) the "planetary" nebulae, small, more or less definite disks or rings each having a hot star, sometimes very faint, at its center (Fig. 61); (2) "extended" or "diffuse" nebulae like the Orion nebula (Fig. 62) which are much larger, with central bright areas which fade gradually into the general sky background. The glowing gas of the planetaries doubtless had its origin in the central star, probably having been

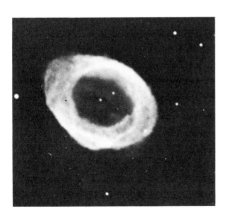

FIG. 61. A well-known planetary nebula, the Ring Nebula in Lyra with its hot central star.

122   *Space Chemistry*

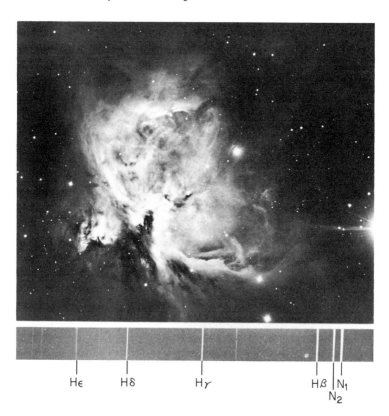

FIG. 62. The Great Nebula in Orion and its spectrum. As in the planetary nebulae the gas is made luminous by the invisible ultraviolet radiation from stars near its center. The lines of the spectrum are the same as those seen in the planetary nebula NGC 6543 (Fig. 60).

thrown out only a few thousand years ago. The extended nebulae may be unusually dense clouds of interstellar gas probably of normal chemical composition. Both types are made luminous by the ultraviolet light coming from the hot stars within them. Hydrogen and oxygen shine most brilliantly.

Spectra of the two types of nebulae do not differ radically, and recent studies have shown that their chemical composition probably is much the same as that of normal stars. The elements (with their atomic numbers) whose spectrum lines actually have been identified in gaseous nebulae are: hydrogen 1*, helium 2*, carbon 6, nitrogen 7*, oxygen 8*, fluorine 9*, neon 10*, magnesium 12, silicon 14, sulfur 16*, chlorine 17*, argon 18*, potassium 19, calcium 20, manganese 25, and iron 26. Measurements of the intensities of spectrum lines have shown that the elements marked by an asterisk have about the same proportions in nebulae as in stars. Proportions of other elements also may be normal but this has not yet been proved.

Between the Stars

Not all the matter in the universe is locked up within stars. Some exists in tenuous form in the chasms between the stars. The stars are compact bodies, really quite large, but almost like points in comparison to the distances between them. (Individual drops of rain 50 miles apart offer an approximate parallel.) Their average densities range from one-millionth that of water for giant red stars to ten times that of water for dwarf red stars. (The "white dwarf" stars are in a class by themselves with very much greater densities.) The density of our sun, 1.4 times that of water, is between these extremes but nearer to the higher one. Between the stars, the density is less by the enormous factor of 10^{24}; there is scarcely one atom per cubic centimeter, but due to the enormous volume it all adds up to a lot of matter. This matter exists in three quite different forms. They are gas, dust, and cosmic rays.

Our now extensive knowledge of the chemistry and motions of matter in interstellar space began near the beginning of the twentieth century when J. Hartmann of Potsdam, Germany, was studying the spectrum of the binary star Delta Orionis. His intention was to study the periodic displacements of the spectrum lines caused by the rapid motion of the star in its tiny orbit—tiny, that is, as seen from the earth, for the star remains fixed as observed directly in a telescope. The spectroscope, however, shows that the star alternately approaches and recedes, as it would if moving in a small, nearly circular, orbit under the gravitational attraction of an unseen companion star. H. Deslandres had previously discovered the interesting changes of motion of this star, and Hartmann set out to measure them more accurately. In so doing, he made an unexpected discovery.

"The calcium line at λ3934 [corresponds to the great K line of calcium in the solar spectrum] exhibits a very peculiar behavior. It is distinguished from all the other lines of this spectrum, first by the fact that it always appears extraordinarily weak, but almost perfectly sharp. [The other lines are wide and diffuse with ill-defined edges.] . . . Closer study . . . led me to the quite surprising result *that the calcium line at λ3934 does not share in the periodic displacements of the lines caused by the orbital motion of the star*" (Fig. 63). The italics are Hartmann's.

Hartmann proceeded to point out that the behavior of this line indicated that it arose, not in the star's atmosphere, but in a detached cloud of calcium vapor somewhere in space on the line joining Delta Orionis to the earth.

The next element to be found between the stars was sodium, which was detected by the behavior in stellar spectra of the double yellow D line. Here as with calcium all the star does is to furnish the background spectrum in which interstellar atoms produce narrow dark lines. Interstellar lines of sodium as well as those of calcium are often quite strong, being readily seen in small-scale spectra (Fig. 64). They have been photographed in the spectra of hundreds of stars.

Stars and Nebulae   125

With large-scale spectra, a few weak lines have shown the presence of three other elements besides calcium and sodium in the gas between the stars, namely, potassium, titanium, and iron. Two kinds of molecules containing carbon also have been found; these are cyanogen, CN, and the hydrocarbon, CH. These molecules in gaseous form are probably not scattered through space like the observed metallic atoms but seem rather to lie somewhere near the

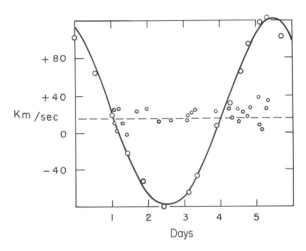

FIG. 63. Velocity curves of the binary star δ Orionis. The interstellar H and K lines of calcium show no certain velocity change.

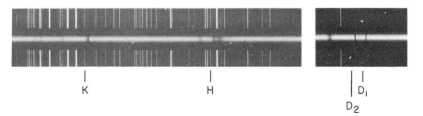

FIG. 64. Interstellar lines of calcium (H and K) and sodium (D_1 and D_2) in the spectrum of χ Aurigae.

stars in whose spectra they are observed. Heat from the stars may possibly evaporate these molecules from interstellar dust.

Plenty of gaseous hydrogen exists between the stars but it does not register in ordinary photographs of stellar spectra. Most of the gas is so tenuous, and collisions between particles are so rare, that nearly all the hydrogen atoms settle down in their normal states where the only lines they can absorb are those in the Lyman series in the far ultraviolet.

The sun's radiation in this ultraviolet region of the spectrum has already been photographed and the time is near when observations will be extended to the stars. Astronomers expect the ultraviolet Lyman lines of interstellar hydrogen to be extremely strong.

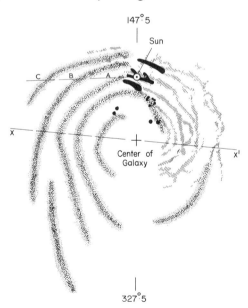

FIG. 65. Spiral structure of our galaxy as indicated by the distribution of hydrogen radiation observed with the 21 cm. line by radio telescopes in recent years. Only the small black areas near the sun were found from ordinary photographic observations.

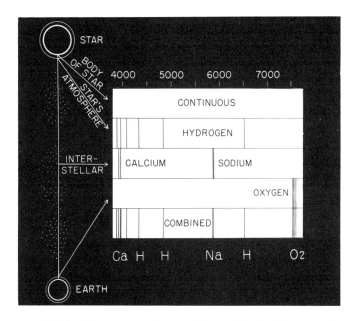

FIG. 66. Possible sources of spectral lines. The dark lines in the spectrum of a star may be formed in the star's atmosphere (represented by hydrogen, H), in the gas between the stars (represented by calcium, Ca, and sodium, Na), or in the earth's atmosphere (represented by oxygen molecules, O_2).

Cool hydrogen atoms can emit radiations of a much longer wavelength than that of ordinary light and have thus recently become observable in the sensational developments of radio astronomy. Observations with huge radio telescopes of the hydrogen radiation having a wavelength of 21 centimeters (8 inches) have verified that hydrogen gas is widespread throughout our Galaxy. Maps of the astronomical distribution of hydrogen are rapidly becoming more accurate and more detailed. They show that the hydrogen in our Galaxy is largely confined to a layer almost as flat as a pancake in which long branches seem to correspond to spiral arms (Fig. 65).

As yet only a few of the more abundant elements have been identified in interstellar gas, but the relative proportions of these seem not to differ markedly from the proportions in typical stars. Moreover, the absence of other cosmically abundant elements can be explained in terms of the physical circumstances involved. Thus the composition of the gas between the stars is probably quite similar to that of the matter locked up in stars.

When an astronomer sees a dark line in a star's spectrum he must decide in which of three possible places it was formed. Was it in the star's own atmosphere, in the gas between the stars, or was it right here in the earth's atmosphere? All three zones produce observable lines. The three possibilities are illustrated schematically in Figure 66 with a few lines typical of each zone. Actually the practical difficulties are minor. With a little experience the relatively few lines from interstellar space and from the earth's atmosphere are easily recognized; the others can safely be assigned to the star.

The presence of finely divided solid particles—we may as well call them dust—between the stars is shown in several different ways. In the first place, V. M. Slipher of the Lowell Observatory and others have shown that surrounding certain stars are faint halos whose light appears similar to that of the star and is almost certainly starlight reflected by small solid particles (Fig. 67). Second, the curious dark markings so common on direct photographs of the heavens must be caused by celestial dust clouds projected against a luminous background (Fig. 68). Third, the light of very distant stars is reddened and polarized as it would be if it had passed through tenuous but extensive clouds of dust.

Concerning the chemical composition of interstellar dust, we have some very fine clues which no one has yet been able to interpret. In the spectra of distant stars near the ground plane of our Galaxy are ten spectral lines (Table 11) whose intensities in various stars increase with

Stars and Nebulae 129

FIG. 67. Bright dust cloud around the star Merope in the Pleiades.

FIG. 68. Dark dust cloud in Orion appearing against a bright background of glowing gas. Sometimes called the Horsehead Nebula.

the amount of reddening. Thus it is virtually certain that these lines are caused by the same dust which produces the reddening. These lines, however, have not been observed in the laboratory and nobody knows to what substance or substances they are due. The reason probably is that the interstellar dust is very cold and that thus far laboratory studies have given relatively little information

TABLE 11

Wavelengths of Absorption Lines Produced by Interstellar Dust

A. Lines slightly widened	B. Lines extremely wide and diffuse
5780.55 A	4430 A
5797.13	4760
6202.99	4890
6269.99	6180
6283.91	
6613.90	

about the spectroscopic properties of substances at very low temperatures. General considerations indicate that the strange lines are due not to rare compounds but to common ones whose spectroscopic properties at low temperatures are not yet known. H. C. van de Hulst of the Leiden Observatory, who has studied the probable molecular combinations of atoms at low temperatures, concludes that the dust grains probably consist largely of the following molecules in a frozen state: water, H_2O; ammonia, NH_3; methane, CH_4; and molecular hydrogen, H_2.

The only other clues we have to the composition of the dust between the stars are rather feeble. One is that when the dust near a hot star is heated by the star's radiation, it seems to give off molecules of hydrocarbon, CH, and of cyanogen, CN. Another hint lies in the polarization of starlight which on its way to the earth has passed through dust clouds. This polarization indicates that the dust particles are somewhat cylindrical and are lined up, presumably by a magnetic field, with their long axes perpendicular to the plane of the Galaxy. This in turn implies the presence in the particles of metal atoms such as those of iron.

A most intriguing possibility is that comets are small clouds of condensed interstellar dust. In 1804 the great French astronomer Laplace suggested that comets might have been formed far outside the known limits of the solar system and then pulled in by the gravitational attraction of the sun. Recent studies by J. H. Oort and G. P. Kuiper confirm this view and place the origin of comets 50,000 or more astronomical units out from the sun. (An astronomical unit is the radius of the earth's orbit about the sun.) That comets consist of typical interstellar materials is by no means certain, but it is plausible enough to deserve further investigation.

Cosmic rays constitute a third form of matter between the stars. In actual mass they are negligible compared to the dust and gas, but they are interesting because of their extremely high velocities and the large amounts of energy they carry. At the earth, and probably at other points along the plane of the Galaxy, it has been estimated that they carry approximately as much energy as do light rays from all the stars. This may not be too impressive if you think of the stars as twinkling luminous points visible only on a dark night. But because of the great volume involved, the whole amount of energy is tremendous.

Cosmic rays normally come to the earth about equally from all directions. Because of collisions with air molecules, virtually none of the original rays reach sea level.

Many fluctuations in the intensity of the rays have been observed but, by and large, the general constancy from day to night and from year to year is remarkable. Such

132 *Space Chemistry*

FIG. 69. The strong solar flare of July 25, 1937, appearing in a spot group near the limb.

fluctuations as do occur—and they are sometimes surprisingly large and rapid—are ascribed to the influence of our sun. Large increases in intensity of low-energy cosmic rays have been observed at the same time as the appearance of a flare on the sun (Fig. 69). On the other hand, charged particles shot out from certain areas of the sun can cause a decrease in cosmic rays on the earth, probably by magnetic deflection of the rays away from the earth. Such a decrease is regularly accompanied by a disturbance of the earth's magnetic field.

Cosmic rays were once believed to consist of radiation of very short wavelengths—hence the term "ray." But that hypothesis was insecure and has been discarded. Today physicists agree that the "rays" are individual atomic nuclei—ordinary atoms with their electrons stripped off—moving with speeds almost equal to that of light. Their chemical composition may be determined by detailed studies of the results of their collisions with atoms and molecules in the air. Hydrogen and helium atoms greatly predominate in cosmic rays and are in very nearly the standard cosmic proportion of about six to one. Carbon, nitrogen, and oxygen may be somewhat more abundant than in stars, but this is not entirely certain. Finally, heavier atoms, notably of iron, cobalt, and nickel, arrive in small numbers, again in nearly the standard proportions. Their extraordinarily high speed makes them observable and interesting.

Because the particles carry positive electric charges (due to the loss of their outer negative electrons) it seems most probable that their high speeds were given to them by passage through interstellar magnetic fields. Future observations from satellites outside the earth's atmosphere and also outside the earth's local magnetic field should bring us closer to solving the difficult problem of the origin and the galactic behavior of cosmic rays.

Galaxies

Among the faint, blurry telescopic patches of light called nebulae by the early observers are some which although rather similar in appearance to the gas clouds previously described are actually totally different. For these objects the surmise that they are distant unresolved groups of stars is correct. Before 1920, astronomers debated whether or not these objects were "island universes." E. P. Hubble settled the argument with the 100-inch telescope on Mount Wilson. His photographs resolved some of the nearer ones into myriads of individual stars and thus gave an affirmative answer to the "island universe" hypothesis. But the

134 *Space Chemistry*

term is an awkward one, and the "universes" are now appropriately called "galaxies" (Figs. 70, 71, and 72). The Milky Way system, our own home galaxy, is undoubtedly typical of many giant galaxies with spiral arms. Within it we have considered in Chapters V and VI the members of our solar system, stars, gaseous nebulae, and interstellar matter. We now proceed to a more general description of galaxies like ours as well as those of somewhat different types.

FIG. 70. A fine spiral galaxy, M51, in Canes Venatici. Our own Milky Way might look like this if seen from a great distance.

FIG. 71. The beautiful spiral galaxy, NGC 6946 Cepheus.

FIG. 72. A spiral galaxy, NGC 4565 Coma Berenices, seen edgewise. The arms lie in a plane. The rounded bulge near the center is the galactic "nucleus." The galaxy is actually composed of billions of stars.

Slight, but probably significant, differences in chemical composition have been noted in stars in various parts of our Galaxy. The stars in the galactic plane, especially those in the spiral arms, have a larger proportion of metals than the stars in the central bulge or "nucleus." The metal-rich stars are now described as belonging to stellar Population I, the others to Population II.

Certain elliptical galaxies without spiral arms seem to contain Population II stars only. Others like our own have both types of stars. Hubble classified galaxies according to their form but he was hesitant to draw conclusions about their chemistry or their evolution. W. W. Morgan is now drawing up a scheme of classification to co-ordinate form with chemistry and evolutionary history. This holds promise of yielding an outline of astronomical evolution on the largest scale yet.

A most extraordinary fact about distant galaxies is that they are all receding from us; and the farther away they are, the faster they are going. Here too chemistry plays a part, for the measurement of velocities depends on the presence of atoms just like those we have on earth. Calcium atoms are most useful, for shifts of the great H and K lines of calcium have given us most of the data on recession of galaxies. In certain galaxies, bright lines emitted by interstellar hydrogen and oxygen have been used.

Observations with radio telescopes of the 21-centimeter hydrogen radiation from several spiral galaxies indicate the presence of a great cloud of hydrogen gas that extends far beyond the photographic limits of the galaxies. These observations are difficult and possibly a bit uncertain but future results are awaited with great interest. Perhaps here we have some real primordial hydrogen gas.

VII. ATOMS, ENERGY, AND EVOLUTION

As we study the present organization of our complex and bewildering, yet orderly, universe and try to think back to its origin and gradual development, we come to two fundamental questions. How were the chemical elements made? What supplies the heat that keeps the stars shining steadily for millions, and even billions of years?

These seem like separate questions, one concerned with matter, the other with energy in the form of heat and light. Indeed for a long time, physicists believed that while matter and energy could interact and change form, they were separate, distinct entities. It seemed that energy could be added to a stationary mass in such a way as to cause it to move but that the total amount of mass—and also the total energy—remained unaltered. Both appeared to be indestructible and uncreatable. This is the old doctrine of the separate "conservation" of matter and of energy. Under this doctrine two interchangeable kinds of energy must be recognized, one inactive or "potential," the other active—such as the "kinetic" energy of an object in motion, or the wave energy of light.

A book lying on a table is pulled downward by the force of gravity but it cannot move because the table holds it up. If pushed off the edge of the table, it falls, gradually losing height but gaining speed. It is really not increasing its total energy but merely exchanging some of its inactive "potential" energy (of height) for active "kinetic" energy (of motion). A powerhouse is said to "generate" energy but really all it does is to change energy from one form to another. In a hydraulic power plant the potential energy of water stored in a high mountain reservoir is converted by gravity into kinetic energy of the same water racing down the flume. At the bottom the water is slowed down by a special rotating device which takes some of the energy and once more changes its form, this time into electrical energy. Similarly if fuel is burned, no energy is created but a part of the energy stored in the atoms is released when they enter into new molecular forms. All this is "classical" physics and chemistry.

In contrast to the doctrine of *separate* "conservation," a new theory has recently been introduced, namely, that matter and energy are actually interchangeable and that only their *sum* is conserved. Matter itself may be considered a form of potential energy. Nuclei of atoms are bits of highly compressed energy which by some strange mechanism we do not yet wholly understand is so tightly locked in that under ordinary circumstances the nuclei persist indefinitely without any change or release of energy. But tap them hard enough and they may react by changing form and releasing some of the stored energy.

Einstein wrote the numerical relationship between mass and energy as follows:

$$E = mc^2$$

which is to say that mass, m, and energy, E, are interchangeable and that when m units of mass disappear they are replaced by E units of energy. The number of energy units is computed by multiplying the mass units by c^2, where c is a numerical constant, actually the speed of light. The speed of light is very great, 186,000 miles per second, and its square, c^2, is tremendous. Thus a little mass can make a lot of energy. A single ounce of matter, if completely converted into energy, would run 400,000 automobiles continuously for 24 hours!

In atomic bombs a small fraction of the original mass is lost either by "fission" (breaking apart heavy atoms) or by "fusion" (forcing light atoms to combine to make heavier ones), but the resulting explosion is terrific, far exceeding that of the bomb where a chemical rearrangement of atoms occurs but where no matter disappears.

Although man has recently learned about this process of producing active energy by releasing some of the potential energy stored in atoms and can now utilize it on a relatively small scale, the sun has used a similar process for billions of years. It is still able to send out so much energy in the form of light—visible and invisible—that only one two-billionth part of that which continuously leaves the sun's surface supplies the earth's needs. (One two-billionth is the fractional part of the whole sky covered by the earth as seen from the sun.)

Astronomers and physicists are convinced that most of the enormous amounts of energy sent out by stars is produced by the internal destruction of a part of their original mass, in the process of making heavy atoms by combining light ones. This idea has guided recent discussions of stellar evolution. In studying the evolutionary history of atoms and stars as well as of plants and animals, it seems natural to assume that all substances, forms, and processes were at first simpler than they are now, and that complexity developed gradually in ordered steps as time passed. The simplest atoms are those of hydrogen. Hence we may postulate that the universe once consisted of huge clouds of cool hydrogen gas, each atom having exactly the same properties hydrogen atoms now possess. All that has happened since is the orderly consequence of those laws of nature now controlling the operation of the universe.

Two major theories have been proposed to account for the formation of chemical elements and the expansion of the universe. One is the "big bang" explanation of George Gamow who believes that the present array of galaxies very distant from each other was preceded by a relatively small, very dense central mass, called "Ylem," composed of protons, neutrons, and electrons. This matter had a temperature of billions of degrees and as it condensed under its own gravity became unstable. Several billion years ago it exploded, sending fragments flying in all directions with a great range of velocities. Chemical elements, at least the lighter ones, helium, lithium, beryllium, and boron, were produced in half an hour; galaxies and stars took longer. As galaxies formed they continued to coast outward, those with the highest velocities now being the farthest out.

In Fred Hoyle's "steady state" hypothesis of continuous creation, the universe had no beginning and will have no end. Individual stars and even galaxies may develop and disappear but the main features of the universe as we see it do not change greatly. If no new matter were being created, the observed outward motion of all distant galaxies would mean that our part of the universe is gradually being denuded of matter. To Hoyle this was an unpleasant idea and he carefully considered what would be necessary to avoid any long-range change in the status quo. As the universe "expands" and some galaxies move out of a certain region, their places would have to be taken by newly formed galaxies. The most likely process to make this possible would be, he thought, the creation of hydrogen atoms at a uniform rate throughout all observable space. He calculated that about 10^{32} tons of matter would have to be created per second to take the place of that which actually seems to be leaving our observable part of the universe. Now this observable part is so large that to accumulate matter at the rate of 10^{32} tons per second, only one hydrogen atom per year would have to be made in a volume of space about equal to that of a large office building. Hydrogen was chosen as the probable original cosmic material, because under suitable circumstances all other elements can be built up out of hydrogen atoms.

Both theories are attractive in certain applications. Gamow's hypothesis provides a geometrically simple and plausible explanation of the expansion of the universe— that is, Hubble's finding that galaxies recede from us with speeds that are in proportion to their distances. Hoyle's ideas lead to the conclusion (which many scientists now believe) that chemical elements have been built out of hydrogen in the interiors of stars. Current discussions, enlivened by rivalry between the two theories, will surely lead to increased comprehension of the origin and evolution of our great universe.

Whatever the history of our universe may have been, scientists agree that we must seek some special source for the enormous stores of energy that the stars have been dispensing for billions of years. Energy can be generated in a star by gradual condensation under its own gravitational attraction. As the atoms fall toward the center they speed up and the gas becomes hotter and hotter. Much energy is developed in this way. But it is not enough; it would serve to keep a star shining for possibly 50,000,000 years or about 1 per cent of its actual life. Ordinary chemical processes such as burning would provide energy to keep the sun shining for a mere 1000 years. The facts seem to indicate that, within a star, energy from some subatomic process is released gradually over a long term of years. The probable process, called nuclear fusion, has recently been intensively studied by astronomers and nuclear physicists.

Let us now outline the probable method by which chemical elements and vast amounts of energy have been produced in the stars now shining, including our sun. Assume

Atoms, Energy, and Evolution   143

that long ago these stars gradually condensed out of huge dark clouds of cool hydrogen. Each one of these protostars would be condensed by its own gravitational pull. As it became smaller and approached the size of a star the central region grew hotter and hotter.

For a while nothing much happened except that more and more atoms became ionized, that is, had their encircling electrons knocked off as collisions of atoms became more frequent and more violent. But as the temperature exceeded 10,000,000° absolute, hydrogen atoms began to fuse together to form helium and some matter was turned into energy.

At these high temperatures all outer electrons are knocked off and the interactions are between bare nuclei. A hydrogen nucleus is a proton with one unit of positive electric charge and one unit of mass. A helium nucleus has two units of positive charge and about four units of mass. Thus it takes four hydrogen nuclei to make one of helium. But it is not a simple case of addition, for two units of positive charge and a small fraction of the original mass

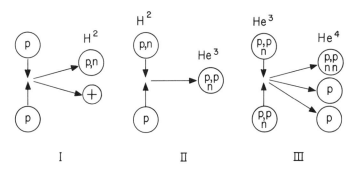

FIG. 73. The three steps in making helium out of hydrogen by fusion.

must be disposed of. Students of atomic physics, bolstered by laboratory experiments, agree that the process requires three successive steps as follows:

I. Two protons colliding at high speed fuse together, emit a positive electron with one unit of charge, and become a nucleus of deuterium which is called an isotope of hydrogen; its nucleus has one proton and one neutron.*
II. A deuterium nucleus fuses with a proton to form nucleus of He^3, a rare but stable form (isotope) of helium.
III. Two nuclei of He^3 fuse to form a nucleus of He^4, ordinary helium, and two free protons.

In symbols these steps are

I. $p + p - e^+ = (p+n) = H^2$
II. $(p+n) + p = (2p+n) = He^3$
III. $(2p+n) + (2p+n) = (2p\ 2n) + 2p = He^4 + 2p$

The graphs in Figure 73 indicate the three steps in making helium out of hydrogen.

Under certain circumstances, another process of making helium out of hydrogen may be more rapid and produce energy faster than the "proton-proton" reaction described above. This is the famous "carbon cycle" first explained by H. A. Bethe. It is a complicated series of six steps involving both carbon and hydrogen atoms, in the course of which four protons are captured and finally come out as a helium nucleus. In the final transformation a carbon nucleus reappears. Thus in a sense carbon acts only as a "catalytic agent" and may serve over and over again. This cycle of helium making requires somewhat higher temperatures

---

*Isotopes of an element are varieties whose atoms have virtually identical electrical and chemical properties but whose nuclei have different numbers of neutrons and thus have different weights. An atom of deuterium weighs twice as much as one of ordinary hydrogen but has the same atomic number, 1, and the same chemical properties.

Atoms, Energy, and Evolution 145

than does the proton-proton reaction, and of course it can operate only if carbon atoms are present.

The proton-proton reaction is believed to predominate and to provide most of the energy in the less massive dwarf stars including our sun, while the carbon cycle may be more important in larger and more massive stars with higher central temperatures.

In any process of making helium out of hydrogen, large quantities of energy are released. This is because the final helium atom weighs a little less than the sum of the weights of the four particles that went into it. The mass that disappears is changed into energy according to Einstein's equation. To keep up the sun's regular output of energy, nearly 600 million tons of hydrogen fuse to form helium every second. The helium thus formed weighs four million tons less than the original hydrogen. The mass that disappears turns into energy which gradually works out from the interior and replaces that lost from the surface by continuous radiation. The quantity of energy released in the interior of a star when hydrogen is converted into helium is far greater than any amount possible by gravitational condensation, and is sufficient to maintain for billions of years the regular output we now observe.

After helium atoms are formed in the interior of a star during its early stages, and as the temperature rises above 100,000,000°, helium and hydrogen nuclei will fuse to make new atoms, chiefly carbon, oxygen, and neon. At temperatures of 1,000,000,000°, magnesium, silicon, sulfur, argon, and calcium will be formed; and at still higher temperatures, iron, nickel, and other metals will appear with the release of a little more energy. Curiously enough in building still heavier elements, energy is *subtracted* rather than added, but the amount is relatively small and causes no appreciable cooling of the gases inside a star.

Thus at last in a single process we have found the answers to both questions asked at the beginning of this chapter. Chemical elements were built out of hot hydrogen gas inside stars. This process, still going on, supplies most of the energy needed by stars in order to keep shining brightly for their lifetimes of billions of years.

Astronomers are convinced that the whole universe is made of the same kinds of atoms we know on earth and in the sun but that they are not everywhere mixed together in exactly the same proportions. The sun appears to represent quite well the average chemical composition of the whole universe; the earth, however, is deficient in hydrogen, helium, and neon. We now know a few groups of stars with exceptional composition. In one of these the metallic elements are much scarcer than in the sun. This discovery has contributed to a new conclusion concerning the evolution of the stars.

FIG. 74. The location of stars of Populations I and II in the great Andromeda Nebula and its companion. In the photograph on the left many of the giant and supergiant stars of Population I are resolved in the outer arms of the spiral. In the hazy region in the upper left corner the Population II stars are not resolved. The companion NGC 205 is composed almost entirely of Population II stars. The extremely bright stars in both photographs are foreground stars.

Atoms, Energy, and Evolution 147

The terms "Population I" for normal stars and "Population II" for stars deficient in metals were introduced about 1944 by the Mount Wilson astronomer Dr. Walter Baade. After extensive, detailed studies of star colors and spectra, he concluded that stars in various galaxies, including our own home galaxy, the Milky Way system, may usefully be classified in two general groups.

This classification is still in the formative stage. Various investigators have adopted different criteria and put the emphasis on different items. Nevertheless, it is undoubtedly of far-reaching importance and is worth discussing because so much of the recent work on stellar evolution has been evoked by Baade's general ideas about Populations I and II, and especially because some simple but most interesting chemical ideas are involved.

Characteristic differences between stars of Populations I and II are listed in Table 12, and are described briefly in the following paragraphs.

TABLE 12
Stars of Populations I and II

	I	II
Location	Disk of galaxy, especially spiral arms	Central nucleus of galaxy; globular star clusters.
Motion	Standard galactic rotation	Special motions
Chemical composition	Metals relatively abundant	Metals scarce
Age	Young	Old

Location. In flat spiral galaxies such as our own Milky Way system, most of the stars that lie near the plane of the system belong to Population I. This is particularly true

148 Space Chemistry

of stars lying in the spiral arms (Fig. 74). On the other hand, stars in the diffuse, nearly spherical nucleus of the Galaxy as well as those in the compact globular star clusters (Fig. 75) belong to Population II. The stars in elliptical galaxies without spiral arms also are of Population II.

Motion. A spiral galaxy is not motionless but is rotating about its center. In our own Galaxy, the sun and most of the nearby stars move in a grand circular orbit at the rate of 140 miles per second. The orbit is so vast that, even at this high speed, to complete a single revolution about 225,000,000 years will be required. Stars of Population I, including our sun, conform to this standard galactic revolution while many stars of Population II do not, but have special individualistic motions. Since we naturally choose our sun as a reference point in measuring stellar velocities, or speeds, we say that stars of Population II have high velocities.

Chemical composition and age. Spectroscopic studies have shown that stars of Population II, especially those in globular star clusters, are poorer in metallic atoms than Population I stars. It is now believed that stars of Population II are the older, being the first to form out of the primordial gas clouds, supposedly composed of pure hydrogen. When, in the course of condensation, central temperatures became high enough, other elements were formed by atomic fusion. Stars of Population I came later and are sometimes called second-generation stars. When they were formed, the residual gas throughout space between the Population II stars was no longer pure hydrogen. It was now contaminated with heavier elements expelled from Population II stars at some phase of their existence.

This contamination could have come about in either or both of two possible ways. It might have arisen from the continuous blowing off of matter from the surfaces of giant red stars. Spectroscopic observations indicate that from the surfaces of the largest cool red stars gases are slowly moving away into space. The rate of discharge is not high

FIG. 75. The globular cluster M92. The stars belong to Population II. Their spectra show very few metallic lines, indicating great age.

but the process is continuous and many stars seem to be taking part in it.

Another more spectacular method of contamination is by explosions of stars. In a typical galaxy about twenty-five stars explode every year. They are called "novae" or new stars, but the term temporary stars would be more descriptive. A previously inconspicuous star suddenly flares up but begins to fade away again in a few days or weeks. Following the outburst, a small cloud of gas appears around the star, at first somewhat resembling a planetary nebula. The cloud expands rapidly and carries out into space a small per cent of the original mass of the star. Once in a few hundred years an even more violent explosion may occur in one of the stars of a galaxy, blowing out into space most of the star's mass. These stars with the more violent explosions, called "supernovae," are at the time of their outbursts the brightest single objects known. For a few days one of them may be as bright as all the other stars in its galaxy put together, but like ordinary novae they quickly fade and again become inconspicuous.

The second-generation Population I stars are believed to form out of this contaminated gas, and thus, even at first, have many atoms heavier than hydrogen. As they condense, additional heavy atoms are built up in their interiors. The development of the more massive stars is relatively rapid, and for a time they shine very brightly; but they exhaust their supply of energy more quickly than Population II stars, lasting for a few hundred million years in contrast to five or ten billion.

The preceding paragraphs give a sketchy outline of current ideas about the development of stars and atoms. The story is not yet definite and complete. At first thought it may seem queer that in this development stars should come first, but many investigators believe that all kinds of atoms except hydrogen were made inside of stars. How the hydrogen itself was formed is a philosophical problem to which astronomical observations do not yield an answer. The difficulty is similar to one recalled by the old question: "If God made the universe, who made God?"

VIII. ROCKETS, SATELLITES, AND SPACE CHEMISTRY

Before long, rockets will soar from the earth to some of the nearer bodies of the solar system and will bring back samples for chemical analysis. First will be those from the moon; next will come samples of the atmosphere and of the solid surface from Venus and Mars. Jupiter is farther away and its surface will be more difficult to explore, but it will eventually be invaded. The sun is so hot that it would vaporize a rocket that came close; fortunately its chemical composition is already well known. The fixed stars are so far away that they are quite out of range and for the foreseeable future we will continue to use spectroscopic methods for their analysis.

In addition to providing transportation to the moon and planets, there is another important way in which rockets and satellites will extend our knowledge of space chemistry. This will be to take instruments aloft, above our atmosphere, where they can record radiations that cannot penetrate the atmosphere. We know that these blocked-off rays carry important chemical information. Already some valuable facts have been obtained from rockets and we look forward hopefully to many more.

For centuries all observations of the stars were visual; astronomers as a matter of course used the same kind of light in their observing that all men use in seeing things on earth. Then photography made it possible to record certain kinds of "light" waves not visible to the eye. First to be used were the ultraviolet rays with wavelengths too short for the eye to see. Later, photographic plates were made that could record infrared rays at the other end of the visible spectrum with wavelengths too long for the eye to see. Now astronomers are beginning to use a still wider range of wavelengths.

To understand what astrophysics has to gain through spectroscopic observations from satellites and from rockets that can fly above most of the earth's atmosphere, we must realize first the great range in wavelength of the radiations that come to us from sun and stars and, second, the severe limitations imposed by the atmosphere on observations made from the ground.

Actually the stars emit the whole gamut of vibrations from hard X rays with wavelengths of a few angstrom units to radio waves hundreds of yards long. (An angstrom unit, A, is one ten-millionth of a millimeter.) In radiations from the stars these extreme wavelengths on either end of the spectrum are quite weak and hard to observe accurately. It will probably be strategic to study first those wavelengths which, in light from the stars, carry the most energy. We already know a great deal about the surface temperatures of various stars, and from the temperature we can calculate which wavelengths will be the strongest. In some stars, like the sun for example, we have already observed them; in others we have not. A brief survey of the situation will help us understand why and how rockets and satellites will be used by astronomers to study the stars.

Rockets, Satellites, and Space Chemistry 153

The position (wavelength) of greatest intensity in a stellar spectrum depends on the temperature of the star's surface. The hotter the surface, the shorter is the wavelength of the strongest radiation. The coolest known stars are red long-period variables at minimum brightness when their surface temperatures are only about 1800° (absolute centigrade). At this temperature the most energetic wavelengths are at 16,000 A in the infrared. The sun has a surface temperature of 5700° and its strongest wavelengths are in the green, about 5000 A. The very hottest stars have surface temperatures of 100,000° with maxima in the far ultraviolet near 300 A.

Table 13 shows how energy is distributed in the spectra of stars with surface temperatures of 2,000° and of 20,000°.

TABLE 13

Percentages of Radiation in Various Regions of the Spectrum

Surface Temperature	Far Ultraviolet 0–3000 A	Near Ultraviolet 3000–4000 A	Visible 4000–7600 A	Infrared 7600 A
2,000°	0.0	0.0	1.4	98.6
20,000°	73.9	11.8	11.3	3.0

A more complete picture of the distribution of energy in light from cool and hot stars is given by Figure 76. In the lower portion of this figure are three curves. The central one, labeled "Eye," shows the sensitivity of the eye to various wavelengths. The long sweeping curve on the right shows the relative intensity of various wavelengths in light from a cool star at a temperature of 2000°. Notice how small a fraction of the energy comes in the visible range. The curve on the left shows the intensities for a hot star at

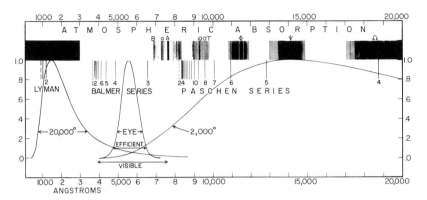

FIG. 76. At the top, regions of the spectrum where atmospheric bands block off light coming in from space beyond are indicated. Next below are the hydrogen series. The lower curves show the wavelengths where the eye is most efficient, and the radiation intensities in the spectrum of sources at temperatures of 20,000° C. and 2,000° C.

a temperature of 20,000°. The maximum energy comes in the blocked-off ultraviolet, and again only a very little is out in the visible range. Still hotter stars are known and with them the case would be still more extreme.

The intensity curve for the sun was omitted in order not to complicate the diagram unduly. It would lie between the two intensity curves just described, with its broad maximum at nearly the same wavelength as that of the narrow maximum of the visibility curve of the eye.

Thus we see that stars send out a great range of wavelengths which we would like to study. But until recently we have been forced to observe the stars from the bottom of an ocean of air many miles deep. In this atmosphere, visible light and the nearby ultraviolet radiation down to about 3000 A come through nicely, and radio waves with wavelengths from 1 centimeter (0.4 inches) to 40 meters (130 feet) also do well. But toward the ultraviolet end of the spectrum, all wavelengths shorter than 2950 A are shut off completely. This limit is set by the absorption of ozone molecules, O_3, having three atoms of oxygen. The ozone layer extends so far above sea level that to receive any large fraction of the ultraviolet light from sun and stars, an optical instrument must be carried up about 30 miles.

Photography may now be applied to the spectral range whose wavelengths are shorter than 14,000 A. Within this range, the upper panel in Figure 76 shows the wavelength regions in which light is obstructed in passage through our atmosphere. The vertical lines in the next lower strip indicate the positions of the stronger lines in the Lyman, Balmer, and Paschen series of hydrogen; the dotted lines show the short wavelength limit of each series. The Balmer series is entirely in the clear and lies in the region where photographic plates are efficient; consequently its behavior in astronomical sources is well known. Absorption bands of water vapor interfere somewhat with the Paschen series in the infrared, but lines in the clear space between 8000 A and 9000 A and also No. 7 at 10,049 A have occasionally been photographed. The important Lyman lines near 1000 A are completely blacked out and can be observed only from rockets or satellites.

Thus far most of our information on space chemistry has been obtained from wavelengths in the range 3000 A–13,000 A, but both shorter and longer wavelengths have made important contributions, especially for the sun. In the future, working lists for stars will be pushed down below 100 A at one end, and above 20,000 A on the other, probably much further for special sources. Wonderful contributions have been made by radio astronomy at much longer wavelengths which are measured not in angstrom units but in inches and yards. Thus far, however, the only definite chemical information obtained from radio waves concerns hydrogen. Most of our knowledge of interstellar hydrogen has been obtained from a narrow radio line having a wavelength of 21 centimeters (8 inches). It is possible that other lines that can yield chemical information may yet be found.

Great success in probing from rockets into the ultraviolet light from the sun has already been obtained, and intensive efforts on both sun and stars are planned for the near future from satellites as well as from rockets.* There is a fundamental reason why astronomical chemists look forward with high expectation to observations of ultraviolet light. In no other region of the spectrum do so many cosmically important elements have their chief lines. Table 14 lists the elements whose strongest lines have wavelengths below 3000 A. Notice how many of the lighter elements it contains; these elements are of predominant interest in astronomical sources, and astrophysicists will get a tremendous boost from being able to observe their chief lines. Estimates of the heights in our atmosphere which must be reached to observe various wavelengths are given in the last line of the table.

The chief lines of the hydrogen atom are those in the famous Lyman series between 1215 A and 912 A in the far ultraviolet. These are the only ones that can be absorbed by cool hydrogen. They will be very strong in the spectra of cool stars. They will be strongly absorbed also by interstellar hydrogen; this may interfere with the observation of these lines in star spectra.

---

*Ultraviolet radiation was discovered by the German physicist Johann Wilhelm Ritter who in 1801 published some novel experiments with sunlight. He showed that silver chloride (the sensitive agent of photographic plates) is affected not only by blue and violet light but by invisible light beyond (of shorter wavelength than) the violet end of the visible spectrum. Credit for the detailed laboratory exploration of this invisible zone is shared by the German physicist V. Schumann and two Americans, T. Lyman and R. A. Millikan. Schumann used the first vacuum spectrograph and made special photographic plates having silver chloride as the sensitive element but with the minimum amount of gelatine (which is nearly opaque to very short wavelengths). His observations extended to wavelength 1200 A, far into the region where air is opaque. Lyman and Millikan finally bridged the gap to X rays of wavelength about 25 A. Lyman's name has been given to the first series of hydrogen whose lines lie between 1215 A and 912 A.

For the element of next greatest abundance, helium, ultraviolet observations will also be valuable. The chief lines absorbed by cool helium gas are between 500 A and 600 A; when this region of the spectrum can be freely observed, our knowledge of cosmic helium will expand rapidly.

TABLE 14
Elements Having Strongest Lines in Ultraviolet Regions

(Limits are in angstrom units)				
500–1000 A	1000–1500 A	1500–2000 A	2000–2500 A	2500–3000 A
He	H	C	Be	Mg
F	N	P	B	Si
Ne	O	S	Zn	Ge
—	Cl	As	Cd	Ta
—	A	Se	Sb	Os
—	Br	I	Te	Ir
—	Kr	Hg	Au	Pt
—	Xe	Rn	Po	—
Altitude needed for observation				
100 miles	75 miles	—	30 miles	—

Observations from rockets will soon cause the element magnesium to become one of the outstanding metals in astronomical studies, probably rivaling calcium in importance. To date our knowledge of its cosmic abundance and behavior has been derived from relatively weak secondary lines. We do know, however, that its atoms are probably about twenty times as numerous as those of calcium. The chief line of the normal, electrically neutral, magnesium atom is in the ultraviolet with a wavelength of 2852 A. This is just below the shortest wavelength transmitted by the atmosphere. When one electron is removed, the atom

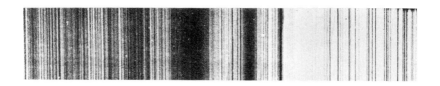

FIG. 77. Lines of magnesium in the sun's spectrum photographed from a rocket. Two lines of ionized magnesium, 2795 A and 2802 A, are blended in a wide dark absorption line; a strong neutral magnesium line appears at 2852 A.

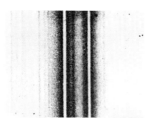

FIG. 78. The pair of ionized magnesium lines near 2800 A photographed from a rocket. The faint emission central cores within the wide absorption lines are barely seen.

is said to be ionized; the chief lines are then a pair at 2795 A and 2802 A (Figs. 77 and 78). It is a favorable circumstance that these lines are so near the neutral line. Comparison of these three lines will aid in determining stellar temperatures, and future observations from rockets and satellites may be strategic in studies of the atmospheres of many kinds of stars.

The first successful attempt to photograph the sun's spectrum from above the ozone layer in the earth's atmosphere was made at White Sands, New Mexico, on October 10, 1946. A special spectrograph designed by scientists of the Naval Research Laboratory was mounted in the tail fin of a German V-2 rocket: a photograph taken at a height of 34 miles with an exposure of 3.6 seconds showed the sun's spectrum to 2100 A. Later attempts pushed the spectrum down further and further toward shorter wavelengths. An early triumph was the photographing of the important ultraviolet lines of magnesium in the sun's spectrum. More recently, a beautiful photograph taken on March 13, 1959, from a Naval Research Laboratory Aerobee rocket at a height of 123 miles extends the spectrum below 600 A; the neutral helium line at 584 A shows plainly (Fig. 79).

Still shorter wavelengths in light from the sun were photographed on June 4, 1958, and March 30, 1959, on Aerobee–Hi rocket flights from Holloman Air Force Base, New Mexico, to heights of about 130 miles. The photographs,

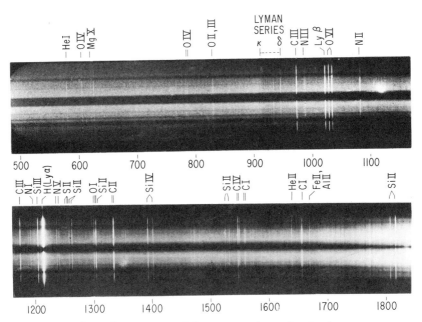

FIG. 79. Ultraviolet spectrum of the sun, 500 A to 1800 A, photographed March 13, 1959, from an Aerobee rocket at a height of 123 miles. These important wavelengths cannot penetrate our atmosphere.

exposed less than a minute, recorded more than 100 bright spectrum lines between wavelengths 1216 A and 84A. The strongest lines were those of hydrogen in the Lyman series, and of helium, 304 A. Other elements identified were oxygen, carbon, nitrogen, and silicon.

The shortest wavelength recorded, 84 A, is less than 1/30 as long as that of the shortest rays of the sun to reach sea level. Thus the range of solar wavelengths in which we can study spectroscopic details has been tremendously extended. In round figures, certain important wavelength limits are as follows:

Shortest wavelength photographed from rockets	80 A
Shortest wavelength to reach sea level	3000 A
Shortest wavelength visible to human eye	3900 A
Longest wavelength visible to human eye	7800 A
Longest wavelength directly on photographic plates*	13000 A

Rocket photographs have presented to us an interesting change in the appearance of the sun's spectrum at about 1700 A. Down to this point, dark (absorption) lines predominate as they do in the parts of the sun's spectrum more usually observed. Near this point, however, there is a switch, with only a little overlapping, to bright (emission) lines. The change has a simple physical explanation: below 1700 A the continuous spectrum becomes so weak that it no longer furnishes a strong enough background to show dark lines. But this weakness of the background makes it easier for bright lines to appear. Bright lines of moderate intensity can here show plainly while similar lines in stronger parts of the spectrum at longer wavelengths would either be completely invisible or would appear as dark lines because the stratum of gas producing them would absorb more light than it emits. Bright lines

*Special photoelectric surfaces can record still longer wavelengths.

will probably appear in the spectra of many normal stars when photographed in the far ultraviolet.

The use of rockets has also expanded the remarkable method of photographing the sun in the light of a single chemical element, described in Chapter V. For hydrogen, many interesting spectroheliograms have been taken with the red H alpha line of the Balmer series, but astronomers

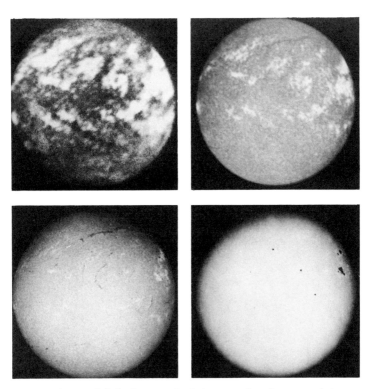

FIG. 80. (Upper left.) Photograph of the sun taken from a rocket at an altitude of about 90 miles using only the strong bright hydrogen Lyman alpha light (1215 A). Three other comparison photographs taken from the ground on the same day are shown. Lower left, Hα; upper right, calcium K (McMath-Hulbert Observatory); lower right, direct photograph in white light.

162 Space Chemistry

long wished to use the much stronger L alpha line of the Lyman series in the far ultraviolet. This was accomplished on the Aerobee rocket flight of March 13, 1959. This exciting pioneer achievement is shown in the upper left corner of Figure 80. Three other photographs of the sun taken from the ground at the same time are included for comparison. The bright patches of glowing hydrogen in the ultraviolet Lyman alpha picture are much stronger than those in the red H alpha picture directly below. Detailed comparisons of these two kinds of photographs will yield much valuable information about the behavior of hydrogen on the sun.

Those and other ultraviolet observations from rockets and satellites offer a most intriguing prospect of extending chemical studies of sun, stars, and nebulae—not in the remote future but in the years immediately ahead.

INDEX

absorption, 5, 15, 24, 43, 46, 67, 111, 160
Adams, W. S., 83
Almagest, 18
aluminum, 39, 57, 66–68, 94
ammonia, 88, 90
Anderson, Carl, 14
Ångström, A. J., 43, 50
angstrom units, 152, 155
antimatter, 14–15
argon, 39–40, 52, 123, 145
astrophysics, 152

Baade, Walter, 147
Balmer series, 46, 155, 161
barium, 40, 113
Berg, O., 57
beryllium, 141
Bethe, H. A., 144
Bohr, Niels, 14, 43–44
boron, 57, 141
Bowen, Ira S., 120
bright lines, 26, 51
Bunsen, Robert Wilhelm von, 5, 23, 32, 43, 55

calcium, 13, 39, 53, 55, 57, 60, 94, 101, 117, 123–25, 137, 145, 157
californium, 60
carbon, 39, 49–50, 52, 60, 82, 91, 113–16, 120, 123, 133, 144–45, 160
carbon compounds, 64–65
carbon cycle, 144–45
Cavendish, Henry, 42
cerium, 116

chemical elements, list of, 36–39
chlorine, 123
chromium, 94
cobalt, 57, 133
color of stars, 30–31, 109, 111
comets, 65, 90–92
Comte, Auguste, 5, 24
Copernicus, Nicolaus, 18
copper, 57
corona, 94–95, 104
cosmic rays, 131–33
curium, 60
cyclotron, 14, 58

dark lines, 23–24, 26, 29–30, 51, 55, 95, 101, 128, 160
Darwin, Sir George, 79
Davy, Sir Humphrey, 55
Deslandres, Henri, 105, 124
deuterium, 58
Draper, Dr. Henry, 29–30
Dunham, T., Jr., 83, 89

earth, 62, 71–74
earth, composition of, 73
Edlen, B., 105
electrons, 14, 16, 26, 35, 40, 43, 51, 55, 141
emission, 5, 15, 24, 44, 46–47, 101, 111, 160
europium, 113

fluorine, 123
Fowler, A., 63
Fraunhofer, Joseph von, 23, 42, 55
Fraunhofer lines, 27

Space Chemistry

gadolinium, 113
galaxies, 6, 13–16, 19, 33, 52, 60, 133–37, 141, 148–49
Galaxy, the, 127–28, 131, 148
Galileo, 19
Gamow, George, 141–42
gravity, 13–14, 16, 70, 74, 90, 131, 139, 142
Greenstein, J. L., 117
Grotrian, W., 105

Hale, G. E., 105
Halley's comet, 90
Hartmann, J., 123–24
Harvard Classification, 31, 108–9
helium, 20, 33–34, 39–40, 46, 48–50, 52–53, 62, 88, 101, 105, 117, 120, 123, 133, 141, 143–46, 157, 159–60
Hipparchus, 18
Hoyle, Fred, 141–42
Hubble, Edwin P., 14, 19, 133, 137, 142
Huggins, Sir William, 27–30, 43, 89, 118, 120–21
Hulst, H. C. van de, 130
hydrogen, 13–16, 29, 34–35, 39, 42–43, 46–47, 50, 52–53, 57, 62, 88–89, 101, 103, 105, 117, 120, 122–23, 126–27, 133, 137, 140–46, 148, 150, 155–56, 160–62
hydrogen lines, 26, 30, 43, 45–46, 89, 108–9
hydrogen line stars, 116
hydrogen molecules, 66–67

inert gases, 52
infrared, 25, 50, 83, 152–53
interstellar dust, 92, 126, 128, 130–31
interstellar gas, 122
interstellar space, 46, 65, 123–26, 128, 131
iron, 29, 31, 39, 56–57, 60, 91, 94, 111, 123, 125, 133, 145

Jansen, P. J. C., 47
Jewell, L. E., 32
Jupiter, 49, 87–89, 151

Kiess, C. C., 84, 89
King, Arthur S., 111
Kirchhoff, Gustav Robert, 5, 23, 27, 32, 43, 55
Kozyrev, N. A., 82
krypton, 40, 52
Kuiper, G. P., 84, 131

lanthanum, 40
Laplace, Pierre Simon de, 131
Lavoisier, A. L., 50
lawrencium, 34
lithium, 34, 40, 53, 141
Lockyer, Sir Norman, 48, 55
lutetium, 40, 116
Lyman, Theodore, 156
Lyman series, 45, 126, 155–56, 160, 162

magnesium, 31, 39, 53, 94, 117, 123, 145, 157, 159
manganese, 57, 59, 94, 123
Mars, 82, 84–86, 89, 151
Masson, A., 43
Maxwell, James Clark, 17
Meggers, W. F., 57–58
Mendeléieff, Dmitri, 40, 57
Mercury, 20, 82
metal oxides, 63–64
meteorites, 39, 57, 75, 77, 92
meteorites, composition of, 77
meteors, 92–94
methane, 88–89
Milky Way system, 6, 13, 15, 19, 118, 134, 147
Millikan, R. A., 156
"Mohole," 72
Mohorovicic, A., 72
"Mohorovicic Discontinuity," 72
molybdenum, 58
moon, 79–82, 89
Morgan, W. W., 137

nebulae, 19, 48, 52, 118–21, 123
nebulium, 33
neodymium, 113
neon, 39–40, 51–52, 60, 62, 123, 145–46
Neptune, 87–89
neptunium, 20

Index

neutrons, 16, 34–35, 60, 62, 141
Newton, Sir Isaac, 13, 20–21
nickel, 39, 57, 91, 94, 133, 145
niobium, 113
nitrogen, 39, 114, 116, 120, 123, 133, 160
Noddack, W. and I., 57
novae, 149
nuclear fusion, 142

Oort, J. H., 131
Orion nebula, 51–52
oxygen, 16, 39, 47, 50–52, 60, 63, 83–84, 113–15, 120–23, 133, 137, 145, 155, 160

Paschen series, 46, 155
"Periodic Table," 40–41, 57
photosphere, 95, 104, 111
Pickering, E. C., 31
plutonium, 20
positrons, 14
potassium, 123, 125
Priestley, Joseph, 50
prominences, 47, 94, 101–3
proton-proton reaction, 144–45
protons, 15–16, 34–35, 62, 141, 143
Ptolemy, 19

quantum theory, 14, 43, 112

radiation, 15, 126–27, 130, 132, 151–54, 156
radio astronomy, 127, 155
radio telescopes, 137
Ramsay, Sir William, 48, 51
red stars, 64, 67, 82, 112–15, 123, 148
reversing layer, 95–96, 100, 104, 111
rhenium, 57, 59
Rowland, Henry A., 32

samarium, 113
Saturn, 87–89
Schumann, V., 156
Secchi, Father Angelo, 30–32, 63, 89, 108

Segrè, Emilio, 14, 58
selenium, 20
Shapley, Harlow, 14, 19
silica, 78
silicon, 39, 57, 94, 123, 145, 160
Sinton, W. M., 84, 86
Slipher, V. M., 128
sodium, 29, 31, 40, 53, 94, 125
sodium, yellow D lines of, 53, 91, 124
solar spectrum, 51–52, 88, 95, 96, 101, 158–59
spectrograph, 21, 101, 158
spectroheliograph, 105
spectroscope, 6–7
spectroscopy, 24
spectrum, 6, 21, 23, 25–27, 29–30, 32, 35, 43, 45
spectrum lines, 24–27, 31–32, 35, 40, 42, 46, 50, 63, 111, 113, 123–24
stellar evolution, 140
stellar spectra, 51–52, 55, 63, 66, 107–10, 112, 114, 124, 126, 153, 156
stellar temperature, 31, 55, 108–9, 111–12, 114, 153, 158
Strong, John, 83
strontium, 39, 113
sulfur, 39, 49, 123, 145
sun, the, 69, 71, 94–95, 105–6
sun, elements in, 98
sunspots, 96
sunspot spectrum, 100
supernova, 60, 149

tantalum, 57
technetium, 57–60, 113
tektites, 77–79
titanium, 27, 31, 111, 114–15, 125
Traverse, Morris W., 51

ultraviolet, 25, 43, 45–46, 50, 53, 122, 126, 152–57, 159, 161–62
uranium, 20
Uranus, 20, 87–89

vanadium, 31
Venus, 82–83, 89, 151

wavelengths, 13, 15, 25, 44–45, 50–51, 63, 66, 127, 132, 152–57, 159–60
white dwarf stars, 117, 123
Wildt, Rupert, 89
Wolf-Rayet stars, 116
Wollaston, W. H., 42

xenon, 40

yttrium, 40, 113

Zeeman effect, 101
zenon, 52
zirconium, 40, 113-15

ILLUSTRATION CREDITS

Figures 3, 5, 7, 12–25, 33–35, 38–45, 47–54, 56, 58–62, 64, 67–72, 74–75.	Mt. Wilson and Palomar Observatories
Figure 9.	From Graff, *Grundriss der Astrophysik*
Figure 28.	By permission, from a painting by Chesley Bonestall
Figures 30–31.	O. Richard Norton, California Academy of Sciences
Figure 36.	R. S. Richardson
Figure 37.	W. M. Sinton
Figure 65.	J. L. Pausey
Figures 77–80.	U.S. Naval Research Laboratory

THE ANTS by Wilhelm Goetsch

THE STARS by W. Kruse and W. Dieckvoss

LIGHT: VISIBLE AND INVISIBLE by Eduard Ruechardt

THE SENSES by Wolfgang von Buddenbrock

EBB AND FLOW: THE TIDES OF EARTH, AIR, AND WATER by Albert Defant

THE BIRDS by Oskar and Katharina Heinroth

PLANET EARTH by Karl Stumpff

ANIMAL CAMOUFLAGE by Adolf Portmann

THE SUN by Karl Kiepenheuer

VIRUS by Wolfhard Weidel

THE EVOLUTION OF MAN by G. H. R. von Koenigswald

SPACE CHEMISTRY by Paul W. Merrill

THE LANGUAGE OF MATHEMATICS by M. Evans Munroe

CRUSTACEANS by Waldo L. Schmitt

TWO-PERSON GAME THEORY: THE ESSENTIAL IDEAS by Anatol Rapoport

THE UNIVERSITY OF MICHIGAN PRESS

DATE DUE

JUL 6 '71			
FEB 21 '72			
OCT 14 '74			
FEB 27 '76			
F			

GAYLORD   PRINTED IN U.S.A.